BUILDING CONSTRUCTION

BUILDING CONSTRUCTION

Service Systems

James Ambrose
University of Southern California

VAN NOSTRAND REINHOLD
New York

Library of Congress Catalog Card Number 91-34671
ISBN 0-442-00294-7

Manufactured in the United States of America

Published by Van Nostrand Reinhold
115 Fifth Avenue
New York, New York 10003

Chapman and Hall
2-6 Boundary Row
London, SE1 8HN, England

Thomas Nelson Australia
102 Dodds Street
South Melbourne 3205
Victoria, Australia

Nelson Canada
1120 Birchmount Road
Scarborough, Ontario M1K 5G4, Canada

16 15 14 13 12 11 10 9 8 7 6 5 4 3 2 1

Library of Congress Cataloging–in–Publication Data

Ambrose, James E.
Building construction : service systems / James Ambrose.
p. cm.
Includes index.
ISBN 0-442-00294-7
1. Building fittings. 2. Buildings—Mechanical systems.
3. Building—Planning. I. Title.
TH6010.A39 1992
690—dc20 91-34671
CIP

Contents

Preface

This book treats a subject that is of interest to a wide range of people with various relationships to the designing and constructing of buildings. There is at any given time a great mass of information on these topics which may be accessed for application to the many tasks of designers, builders, suppliers, and others. The enormity of this information resource is at once reassuring to those who regularly encounter needs for it, and overwhelming to just about everyone who needs to figure out how to use it.

This book relates the topic of building construction to the basic problems of incorporation of various building services and the construction requirements for their development. The basic assumption here is that the need to design precedes the need to build, and that real concern for how to build comes from a desire to build something. This is the normal process of development for designers and others who start from the point of desiring a building and then proceed to determine what it should be. Intense concern for specific consideration of building materials, systems, and details of construction thus emerges at a later stage of design, typically after the general form, size, and essential nature of the building are already proposed.

A major problem with using the great mass of available information about building construction is that it is largely not oriented to the purposes of education or design; thus it is not user-friendly to the student or apprentice in architecture, or to any others who do not already have a broad grasp of what buildings and building construction are all about. Rich as they may be as information sources, *Sweets Catalog Files*, the *CSI Spec-Data System*, and *Architectural Graphic Standards* are not friendly to inexperienced users. You have to already know the trade lingo, what you are looking for, and pretty much why you are looking for it to make effective use of these resources.

The latest, most extensive, most detailed information about building construction is largely produced by persons engaged in the manufacturing, supplying, specifying, purchasing, regulating, financing, or insuring of building materials and products, and in their applications to making buildings. The information presented is often slanted toward the specific concerns of those who produce it. Thus it is to be expected that materials forthcoming from manufacturers and suppliers are shaped to the purpose of promoting sales of their products, that insurers and financiers have strong economic concerns, and that specification writers and enforcers of building codes are somewhat paranoid about specific language and terminology that is legally binding.

This book attempts to be user-friendly to the person who is basically more interested in buildings in general and somewhat less in the specific concerns for their construction. The basic purpose here is to develop a general view of buildings and the problems of their design as they relate to the eventual need to construct them. Highly specific, detailed information, such as how to attach gypsum drywall to wood studs, can be pursued through appropriate sources, once the real need for the information is established. The reasons for having the drywall, for attaching it to the studs, for having wood studs—and indeed, for having the wall in the first place—need to precede the search for the specific information.

Buildings are complex and the topic of their construction is correspondingly extensive. In order to cut down the size of this presentation, the topic here is limited to concerns for the development of the various building services, including those for power, lighting, communication, water supply, gas, waste removal, roof drainage, fire detection and fighting, and vertical transportation. As extensive as those concerns may be, it leaves a lot of the building not treated. This book is, in fact, the fourth volume of a series. The first volume treats the subject of the building's basic shell that forms the enclosure; consisting primarily of roofs, exterior walls, windows, and doors. The second volume treats the subject of the building interior; consisting primarily of floors, interior walls, ceilings, and stairs. The third volume treats the problems of site and below-grade construction.

My views of the need for this book and my ideas for its contents have emerged from many years of experience as a building designer and a teacher. I am grateful to all my former students, coworkers, and others whose feedback of frustrations have helped to shape those views and ideas. I am also grateful to the many people at Van Nostrand Reinhold who have helped to bring my rough materials into being as an actual book; particularly my editors, Everett Smethurst and Wendy Lochner, as well as Cindy Zigmund and Alberta Gordon.

I am also grateful to various organizations who have permitted the use of materials from their publications, as noted throughout the book. It should be noted that, while I have striven to follow manufacturers' recommendations, the details for equipment installation shown here are my own interpretations and should not be blamed on the suppliers of information that I may inadvertently have misused in some instances.

Finally, I am grateful to my family, who support and tolerate me in my home office working situation. For this book, I am particularly grateful for the assistance of my wife, Peggy, and my son, Jeff.

James Ambrose

BUILDING CONSTRUCTION

1

Introduction

Design for the development of various building services is a major area of concern in the general design of buildings. This chapter presents discussions of the general issues of building services and their relations to the development of the building construction and how the topic is developed in this book.

1.1 DESIGN FOR SERVICES: GENERAL CONSIDERATIONS

This book deals with the construction of buildings and with the considerations that must be made for accommodation of the building service systems. The general problem of building services has many aspects, ordinarily involving the concerns of many designers (see Figure 1.1). The performance of service systems is of direct concern to the specialists who design the individual systems, but it is also a general concern for the architect. Of particular concern to the architect are the problems of incorporating the systems into the building construction, especially as it regards highly visible elements or those requiring considerable space in the building.

A major concern for the architect is the blending of all individual service systems into the whole fabric of the building, an integrative task that is usually a major responsibility for the architect. This requires that the individual designers supply the architect with complete information, but also typically means that the architect must *anticipate* the needs of the individual systems and not be caught by surprise after the design work is completed.

The primary topic of this book is not service systems themselves, but rather their basic construction considerations. This has a twofold meaning, the first relating to the specific details required for the installation and proper functioning of the elements of the service systems. However, the selection of materials, components, systems, and details for the general building construction and the general planning of the building often impinges on the performance requirements for various service systems. Thus the design of the whole building should preferably be a fully interactive, integrated process.

While it is possible to consider only a few concerns, or even a single one at a single point in the design work, all of the potential effects of choices for the construction must eventually be dealt with. This is hard to achieve in reality, but the inseparable nature of the work must be understood and dealt with in some fashion. In the end, the whole building must be designed, even though specific situations may be dealt with as individual problems.

For purposes of illustration in this book, a particular form of construction is often

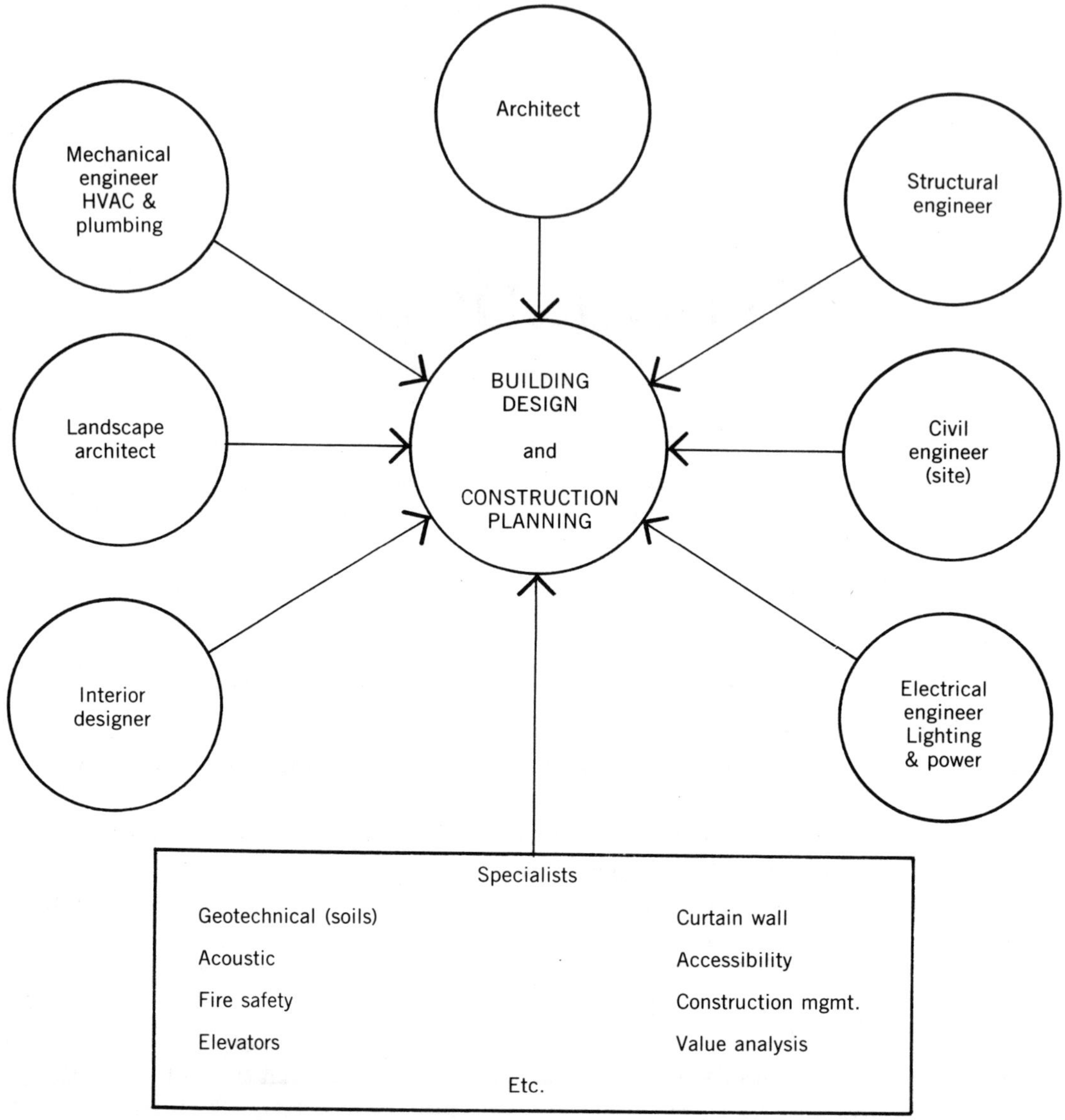

Figure 1.1 Interaction of the designers for the building service systems.

presented and the problems of relating some service elements to it described. This should not be taken to imply any preferred order of decision making: it is simply a practical device for illustration. Anticipating needs for various services may well inhibit choices for various forms of construction, even though we seldom start by designing service systems first.

The point of view here is that of the designer and what is of prime concern in the design process. The writing of specifications, management of construction, and production of materials and components for construction are necessary activities that will be considered, but design is the focus.

1.2 BUILDING CONSTRUCTION AS A DESIGN PROBLEM

The making of buildings is a big business, affecting many people. However, the development of the materials in this book is based, in part, on the attitude that the determination of building construction is an architectural design problem.

The solving of problems, large and small, with regard to the construction work affects the design—or at least it ought to. If certain details or dimensions are not feasible to achieve with a particular material or system, the building designers should accept the facts, or prove by some means that what is proposed for the design can truly be accomplished. Design work must be informed, not necessarily in early sketching stages, but as soon as any serious commitments are made and the detailed development of the work proceeds.

It is possible, of course, to make the building construction itself a major design determinant. However, respect for correctness and expression of the construction is one thing; reverence for it is another. Many architects are more concerned with form, space, appearance, illusion, metaphor, symbolic relations, human response, or other matters, caring somewhat less about the expression of the construction or the revealing of the building's functioning elements.

For most designers the actual work of design usually is a very personal matter. Launching points, decision sequences, and the relative weights of different values vary considerably from designer to designer as well as from project to project with a single designer. Approaching the subject of building construction as an architectural design issue simply means that concerns for the construction emerge as necessities in the process of design. The designer does not set out to produce some construction, but rather encounters construction considerations in the normal process of designing a building. First comes the concept of the building and all of its planning concerns—then the problems of making it.

There is no intention to advocate any particular design philosophy in this book. Proper construction is considered essentially as a simple, pragmatic matter. It is considered to have a reasonable value and to be something that deserves serious attention—but it is not essential that it command dominance in all architectural design decisions.

1.3 HOW TO USE THIS BOOK

The basic purpose of this book is to help the inexperienced designer, whether in a school design class or in an apprentice position in a design office. However, anyone who is interested in the topic of building construction and its effects on building design should be able to derive some use from this book.

There is no single method for the most effective use of the materials in this book. Readers are sure to vary considerably with regard to interests, needs, and backgrounds. While the materials have been arranged with some logical sequence in mind, it is anticipated that the typical reader will take up the materials in a random fashion. This is to be expected all the more if the reader is seeking some help in the progress of some design work.

We do not mean to discourage any reader from taking up the book materials in the sequence presented. This is likely to make the most sense if the reader is essentially uninformed in the general topic. Development of a fluent "construction" vocabulary is also to be generally encouraged, as the discipline of proper terminology is quite essential for accurate communication in the building design and construction business world.

Finally, for many readers, this book may serve a principal function in pointing out sources for additional information. General major sources are discussed in the Sect. 1.4, and various special sources are cited throughout the book in the discussions of specific topics.

1.4 SOURCES OF DESIGN INFORMATION

Information about building construction is forthcoming from many sources and relates to many potential uses (see Figure 1.2). For the specific purpose of supporting design activity, some particular sources and applications are of special value. This is a highly variable situation, responding to many considerations for a particular design activity. Significant variables include the following:

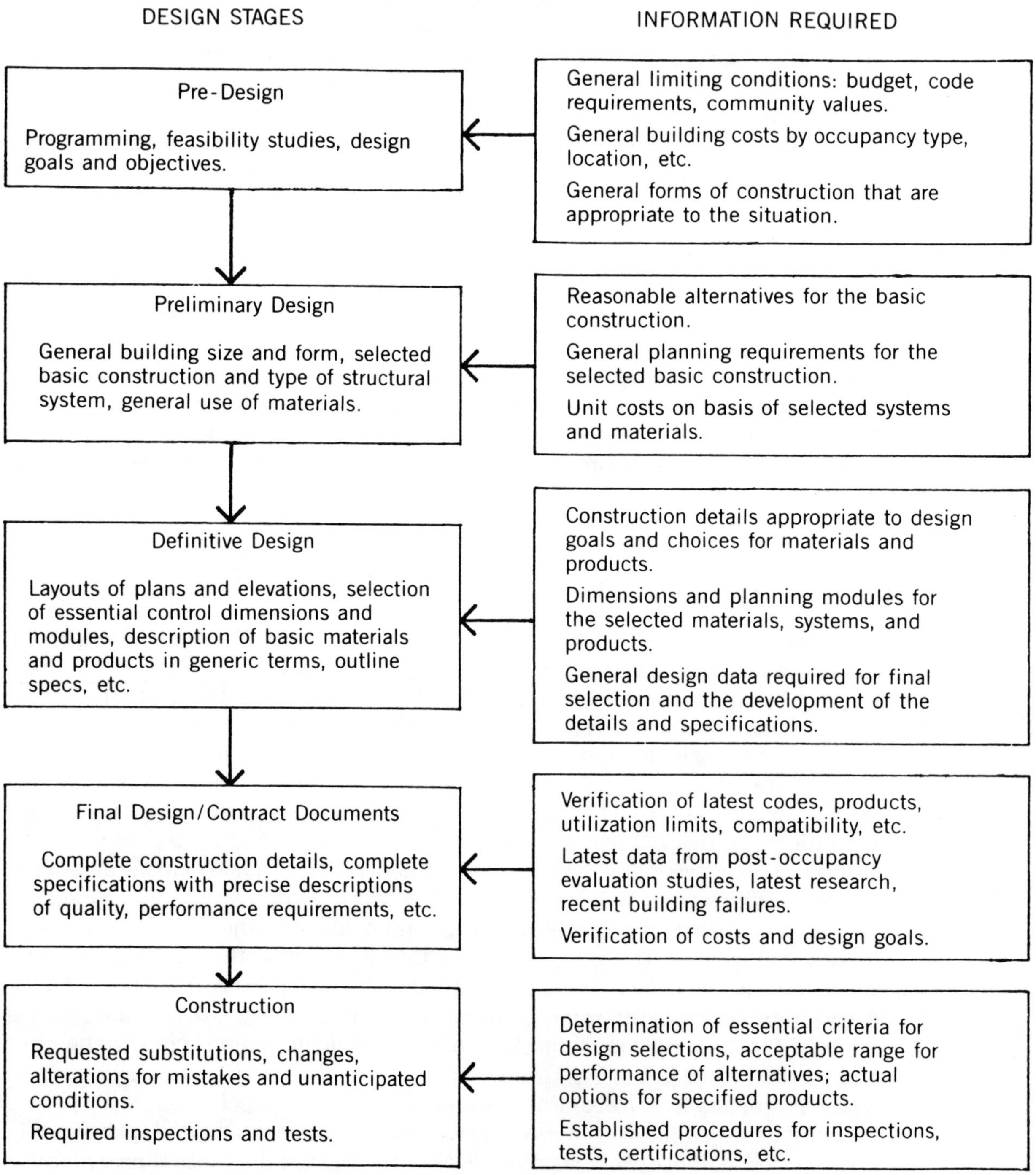

Figure 1.2 Use of construction information in the building design process.

Stage of the design process—from very preliminary to final construction directives

Nature of the design project—large or small; luxury or low cost; specific constraints (codes, special occupant problems, etc.)

Working context of the designer—time constraints; design budget; size of the design team; access to information sources

Individual types of information and types of information sources may be of particular value, depending on the design work as qualified by the preceding, as well as by many additional considerations. Some major categories of information sources that are generally supportive of design work are the following:

Books—including ones of broad scope, such as general references on building construction, and ones of a highly specialized nature with important data for a very specific problem

Industry-supplied materials—most supplied for promotional purposes, but nevertheless usually indispensable for specific product data

Records of designs—statistical, graphic, or photographic records of proposed or actual construction

For any specific design work, the truly usable or necessary references may relate as much to the particular needs of the individual designer as to the specific nature of the design activity. The education, training, developed skills, and previous experience of the designer may well make certain general references indispensable or generally require help only for very special situations.

This book is not intended as an inexhaustible source of specific information about all of building construction. Although the view is broad, the scope of the presentations here is narrowly bound by specific concentration on the needs of the designer, the general nature of design work, and a particular interest in the facilitation of building services. It is generally assumed that the reader, whether in a school or working in a design office, has access to some general information sources. These incluce the following basic items:

One or more general texts on basic building construction materials, products, and processes

One or more general references on standard details of building construction

Some industry-supplied information; as generally represented by *Sweets Catalog Files* or a selected, personally assembled set of individual entires as obtainable from individual manufacturers or agencies

It is assumed that perceptive readers are aware of some of the pitfalls in using information sources. We refer particularly to the major concerns for timeliness, neutrality of the suppliers of the information, and feasible applicability to the design work at hand. In the latter regard are such concerns as local codes, availability of products and services, climate differences, and so on.

Building construction is largely achieved with commercially developed products. The suppliers of these products do a lot of hard selling to compete in the marketplace, and information obtained from them is quite understandably not of a neutral, objective nature. You can't expect to be able to judge the relative appropriateness of a product to a specific application on the basis of the information supplied by the product's manufacturer or supplier.

Many references have been used in the development of the work presented in this book. Some of these sources are cited for very specific information, but the general development of the work derives often from a collective consideration of many sources. For general reference, we acknowledge—and recommend to the reader—the following sources:

1. *Architectural Graphic Standards,* 8th edition, by Ramsey and Sleeper, Wiley, 1988.
2. *Fundamentals of Building Construction: Materials and Methods,* 2nd edition, by Edward Allen, Wiley, 1990.
3. *Construction Materials and Processes,* 3rd edition, by Don Watson, McGraw-Hill, 1986.
4. *Construction Principles, Materials, and Methods,* 5th edition, by Olin, Schmidt, and Lewis, Van Nostrand Reinhold, 1983.
5. *Sweet's Catalog Files: Products for General Building and Renovation,* annually published, McGraw-Hill.
6. *Mechanical and Electrical Equipment for Buildings,* 7th edition, by Stein, Reynolds, and McGuinness, Wiley, 1986.

For specific problems and applications, many additional references are cited throughout the book. While timeliness of any source should be noted, it may be less significant than the general overview obtained from a well-organized general reference prepared by essentially neutral authors.

2

Architectural Components

This chapter presents materials relating to the most ordinary and generally indispensable parts that occur in the majority of buildings and with some of the design concerns that arise in producing them. While designers are necessarily concerned with all of a building's parts, the concentration here is on those parts most directly involved in relation with the various building services, which comprise the central topic of this book.

Any selected building part must be designed in the context of the problems of the whole building. Dealing with the individual elements of building construction requires some attention to their eventual coordination in the whole constructed building. Putting the parts together is the theme of the work in Chapter 5 of this book, where the whole construction of several buildings is presented. The objective in this chapter is to deal generally with the various concerns for the basic building component parts.

2.1 INDIVIDUAL BUILDING COMPONENTS

For the purpose of discussion, we will consider a two-level set of basic architectural components. On the first level are those major elements that principally determine the structure and basic character of the construction of the building. As shown in Figure 2.1, these major elements are the roof, floors, exterior walls, foundation elements, and elements for vertical circulation (stairs, elevators, etc.).

On the second level are elements which, although necessary, have generally less influence on determination of the basic form of the structure or the choice of the principal materials and methods of construction. The second level includes windows, doors, ceilings, and various items involved in the building service systems (for power, lighting, HVAC, etc.). The second level may also contain site features outside the building.

In any given situation one or more of the second-level elemenets may take a dominant role in the set of values for design. For example, in an office building, the need for development of an integrated ceiling system may strongly constrain the design of the roof of floor structure, interior partitioning, and services for lighting and ceiling-mounted HVAC elements, fire sprinklers, intercom systems, and so on.

While the general purpose of this book is to deal with building construction, the

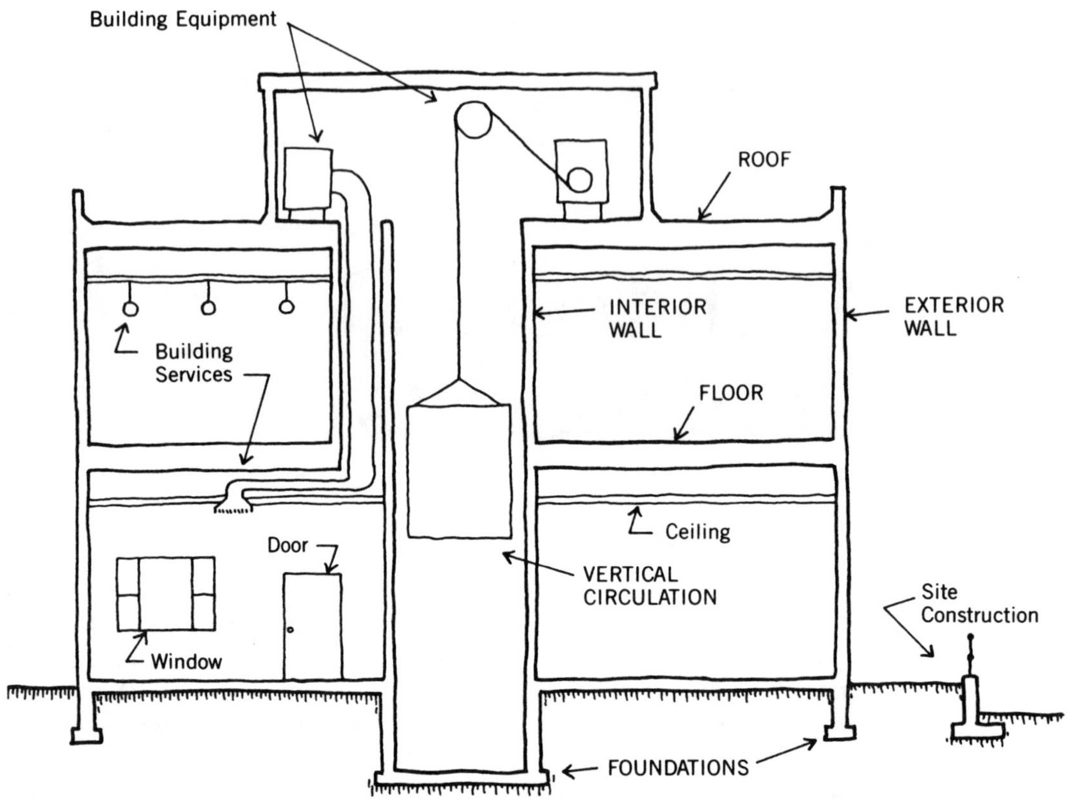

Figure 2.1 Fundamental construction components for buildings.

specific issues for this volume are the problems of construction related to the development of the building service systems. These systems must be seen in their context in the whole building, but they must also be individually understood. The remaining sections of this chaper deal with the major, common individual service systems in buildings, the intent being to explain them in terms of their basic requirements and general forms.

The discussions in this chapter are essentially limited to the fundamental concerns for the service subsystems themselves. It is important to have some grasp of these basic concerns, although the actual design of these subsystems is not primarily what this book is about. Chapter 3 deals more extensively with the more critical concerns for the designer of the general building construction in terms of the relation between construction and the service systems.

2.2 HVAC SYSTEMS

Heating, ventilating (air freshening), cooling, and the general treatment of the building's interior thermal and atmospheric condition is a multidimensioned task. This is often lumped under the general activity termed HVAC, which stands for heating, ventilating, and air conditioning.

By itself, the term *air conditioning* is often taken to mean essentially the cooling of interior air. However, even in this limited context, it typically also involves the removal of moisture from the air. As referred to in the HVAC term, air conditioning is meant to stand for the complete conditioning of the air, which may include the filtering out of dust and odors, freshening with outdoor air, adjustment (up or down) of the temperature, and adjustment (up or down) of the relative humidity.

The general purpose of having an HVAC system is to maintain some level of healthful and comfortable conditions for people inside a building (see Figure 2.2). This problem has many variables, including the following major ones.

Local climate: The *need* to heat or cool, or to add or remove moisture in the air, depends on the local climate, as well as the season of the year.

Building occupancy: Who is occupying the building (athletes, hospital patients, ect.) and what they are expected to do in the building (sleep, perform aerobics, etc.) makes a difference for what is an acceptable interior condition.

Building size, shape, construction type, etc.: This affects what form of system may be used and how the operating equipment may interact with the general building construction.

HVAC systems generally include the following basic components:

A heat-generating system (furnace, fireplace, etc.)

A cooling system (refrigeration unit, evaporative cooler, etc.)

An air-handling system (circulating, exhausting, refreshing, etc.)

A control system for hand adjusting and/or automatic monitoring of the system operation

Functions can be handled individually or combined in various ways. In a so-called all-air system the basic heating or cooling may be achieved as part of a whole process

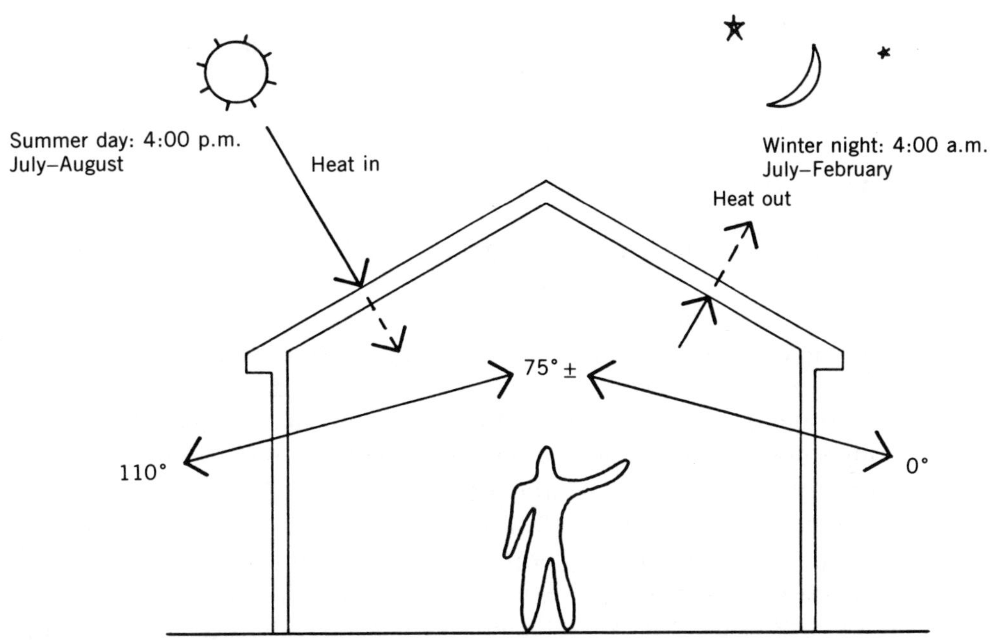

Figure 2.2 Fundamental concerns for human comfort in building interiors.

of conditioning the air. Cooling is most often achieved this way, although other heating methods are sometimes more effective.

Major operating concerns are for the control of the system and the amount of energy consumed. Cooling in steamy climates and heating in frigid ones are major sources for energy consumption and typically receive a lot of attention in terms of achieving optimal designs for energy savings.

From an economic point of view, the building construction represents primarily a major, one-time cost investment. Initially installed HVAC systems are similar, but the more critical concern may be for their operating costs over some period of time. If modifications of the building construction can reduce these long-term costs, they may justify some higher initial investments.

Other factors that affect the relations between HVAC systems and the general building design include the following.

Space for HVAC equipment: Space must be allotted for equipment and it must often be strategically located for its feasible operation.

Noise and vibration: These are critical factors for cooling equipment and large fans. Equipment must also be accessible for maintenance and replacement.

Space for air duct systems: These must serve the building's interior spaces and link up with the operating equipment. Air intake and exhaust must connect to outdoor air at appropriate locations.

Properties of the building enclosure: Thermal responses and air leakage of roofs and exterior walls will affect the task required of the HVAC system.

Building planning: Various factors of the buildinng form and interior arrangement will establish the nature of the HVAC problems. Compass orientation for sun exposure, locations and numbers of windows, number of discrete interior spaces, and many other factors will affect the complexity and efficiency of the HVAC system and the choices for basic types of systems.

For various purposes, it is advisable to closely integrate the designs of the HVAC and lighting systems for a building. Various aspects of this are discussed in Chapter 4 with regard to the development of integrated systems.

2.3 LIGHTING SYSTEMS

Lighting (actually illumination) is usually achieved in buildings by some combination of daylight and electrical lighting fixtures (see Figure 2.3). During daylight hours the building site and general exterior, as well as some parts of the interior, are typically illuminated by daylight. In these energy-conscious times, it is usually a design goal to make *maximum* use of daylight for energy and cost savings.

At night, as well as deep inside the building at any time, some other form of lighting must be provided, typically from electrically powered lighting systems. This involves the installation and maintenance of the light fixtures, installation and occasional maintenance or rewiring of the electrical power distribution system, and some means of control of the system (switches, dimmers, etc.).

For large commercial and institutional buildings, especially those having long hours of usage, the powered lighting system may be a major user of energy. In these cases, striving for an efficient lighting system may become a major design goal for the

Figure 2.3 Interaction of natural and artificial illumination for buildings during daylight hours.

building. While this affects the design of the lighting system itself, it also quite often relates to the general planning of the building to reduce the demands for lighting.

In reducing the need for powered lighting, the least flexible item is the need of the building user. User need may involve the necessity to perform some specific act (walk down a stair, read a book, etc.), usually defined as the visual task. Or it may relate to some goal, such as the proper viewing of a displayed sales item. These are the things that need doing, and the whole lighting system must respond. Proper selection of lighting equipment and proper installation in terms of positioning units may permit the powered system to do an optimal job of providing the necessary illumination.

The planning and detailing of the building form and its construction have a considerable effect on defining the illumination problems. Selection of internal finish materials in particular can have a major effect on the operation of the internal illumination systems, as can ceiling heights, room height-to-width ratios, and the number, height, size and placement of windows.

In warmer climates, where solar heating is a major thermal comfort problem, the design for use of daylight must be very carefully integrated with that for solar shading from heat gains. And now, in any climate, it is generally desirable to have close coordination in the general designs for HVAC concerns and lighting, as part of an overall energy use reduction effort.

Lighting presents many technical problems, requiring good engineering analysis and design. However, illumination is also a major architectural design problem: If the building can't be seen, how can it be appreciated architecturally? Therefore, for purely *visual* reasons, the architect must be sensitive to the issues and solutions for lighting problems. Equally important, the implications of choices of building form and detail with respect to influence on lighting problems must be appreciated.

2.4 POWER SUPPLY SYSTEMS

The major power source for buildings is electricity, typically generated somewhere off the building site and delivered in wired form to the building (see Figure 2.4). This *power* source is actually an *energy* source with a potential for many uses, some of which can be done by other sources, such as gas, oil, steam, hydraulic force, or the sun. But to operate the artificial illumination system, run the computers, make the fans blow, and operate numerous appliances, we need electricity.

The electrical supply system is permanently attached (hard-wired) to many elements, such as the HVAC equipment, lighting fixtures, elevators, and so on. However, it must also supply an open plug-in system for the more or less random attachment of portable items, such as TVs, desk lamps, power tools, and kitchen hand appliances.

Installation of the electrical power system usually has only a minor influence on architectural design or construction development issues. There are many special requirements for the details of the electrical systems themselves, but only a few that affect the general construction; some of these are discussed in detail in Chapter 3.

A major concern for electricity is the potential danger for fire or for severe electrical shock. Codes have many requirements for the systems that originate as concerns for safety. In some situations the building construction is relied on to protect occupants from the electrical system, where elements are buried inside the construction. Today, however, the systems themselves are mostly made as user-safe as possible on their own.

For large buildings there are many pieces of equipment that must be housed. This involves both a space commitment and some attention to the security of the equipment—providing protection against misuse and protecting occupants from accidental dangerous shocks.

With our ever-expanding technology, demands for use of electricity often increase after initial construction of a building or as user needs change. Access to the electrical equipment and wiring must be reasonable to permit alterations of the system.

Figure 2.4 Hookup of a building to an off-site source for electrical power.

2.5 SIGNAL AND COMMUNICATION SYSTEMS

Just about every residence these days has a doorbell, a telephone system, and a thermostatic control for its HVAC system. Each of these systems is typically operated with a separately wired, low-voltage, direct-current electrical system. In many buildings there are also cable TV systems, computer networks, public address or intercom systems, and fire alarm systems, all using their own wired systems.

Many of these wired systems may be built into the construction, in the same general manner as the basic electrical power system, although safety requirements for low-voltage systems are considerably eased. Depending on the degree of permanence of the systems, however, they may be somewhat less *hard-wired;* that is, they may edge over to the class of systems more similar to those typically installed by building users for their personal hi-fi systems or small-scale computer networks.

Some issues regarding these sytems are discussed in Sec. 3.4.4 and in the building case studies in Chapter 5.

2.6 PIPED SYSTEMS

The most common piped systems are those installed for the water supply, waste drainage, and natural gas supply systems (see Figure 2.6). Each of these systems has special concerns, for example, pressurization in the water supply and gas supply

Figure 2.5 Elements of interior wired systems for communication and signalling: Modular phone jacks.

systems, and gravity flow for drainage in the waste systems. These systems also typically connect to off-site mains, presenting problems for the site design and the general building base and foundation, as well as details for incorporation into the building construction.

Piping may also be required for fire sprinklers or other fire-fighting purposes, for irrigation, for internal roof drainage (usually separate from the building's sanitary waste system from toilets, etc.), and sometimes for pressurized air or other special liquids or gases in hospitals or laboratories.

Piped systems usually have three components: a supply point (delivered from the mains, for example), the piped distribution or collection system, and terminals at some form of fixture (toilet, sink, faucet, etc.). The building entry from the mains, the piping network, and the fixtures must all be accommodated in the construction. Concerns for the various common systems are discussed in Sec. 3.2 and 3.3, and some specific situations are illustrated in the building case studies in Chapter 5.

Piping is also used with most HVAC systems, and some of the special problems for this are discussed in Sec. 3.8 and 3.9.

2.7 VERTICAL TRANSPORTATION SYSTEMS

Buildings with multiple levels require means for movement of people and materials from level to level. Movement may be achieved with ramps and stairs in some cases,

(a)

(b)

Figure 2.6 Piped systems for building services: (a) Water supply. (b) Natural gas supply.

but various powered systems may also be employed (see Figure 2.7). The most common of these systems are elevators and escalators. In many commercial and industrial buildings, however, there are often special systems for movement of materials or equipment.

Vertical transportation systems take up considerable interior space in buildings. This space begins with that required for the housing of the traveling carrier (elevator cab, belt loop of the escalator, etc.), but also includes housing for operating equipment and the usual overruns at the ends of the travel path. In addition to the space-planning problems, there are often concerns for the noise and vibrations of the equipment and the moving of transporting elements.

General problems of transportation systems are discussed in Sec. 3.7. Some details for installation of elevators are shown for buildings 4 and 7 in Chapter 5.

Figure 2.7 An open cab elevator, installed in an atrium space in a multistory building.

2.8 SOUND CONTROL SYSTEMS

Control of sound in buildings is not so much done with a "system" as it is with special attention to the form and details of the construction. Design for sound control begins with an analysis of the potential problems (maintaining privacy, containing noisy sound sources, enhancing acoustic conditions in single spaces, etc.). Solutions for these may typically involve combinations of efforts: planning for separation, choice of surfacing materials, shaping of interior forms, and modifications of the construction of separating elements (walls, roofs, floors, etc.).

Sound may be carried by the air or be transmitted through the solid elements of the

Figure 2.8 Built-in elements of a fire-fighting system: glass-fronted cabinet with fire extinguisher.

building construction. Both transmission paths need consideration and involve different properties and details of the construction.

Some special concerns for sound are discussed in Sec. 3.5, and some particular situations are illustrated in the case studies in Chapter 5.

2.9 FIRE CONTROL SYSTEMS

Fires in buildings are best "controlled" by eliminating the hazard for their ignition and spread in the first place. This can only be partly controlled by the building designers; building users can create hazardous conditions by moving dangerous or simply highly flammable materials into any building, by overloading electrical equipment or circuits, by carelessly using matches or smoking materials, and so on.

It must therefore be assumed that a fire can occur in *any* building, and the scenario for fighting the fire and for safely removing the building's occupants must be played out. A major portion of building code requirements is devoted to these concerns, and no other single concern has a greater impact on the restriction of use of the materials, form, and details of building construction.

In addition to the determination of the basic construction, there are often special elements and systems required for the total fire control, detection, alarm, and suppression (see Figure 2.8). These elements may include lighted exit signs, automatic door closers, water sprinklers, fire hose cabinets, special water supply for fire fighters, smoke alarms, and so on. The total fire control "system" thus virtually involves the *whole* building.

Some general concerns for elements of fire control systems are discussed in Sec. 3.6. Some special construction details relating to fire control are illustrated in the case studies in Chapter 5.

3

Construction Development of Service Systems

This chapter deals with the various service systems, regarding their basic elements, essential functions, and the specific issues that relate to their incorporation into the building construction. Some general system design considerations are treated in Chapter 4. Specific examples of systems are illustrated in Chapter 5 for nine case study buildings.

3.1 AIR-HANDLING SYSTEMS

Air is necessary for humans, and wherever there are spaces in buildings that are intended for human occupancy for some period of time (not just walking through, etc.), it is generally necessary to maintain a supply of air. Air will generally flow into any space that is not fully sealed off, but the issue usually becomes one of supplying some "fresh" air, that is, the kind of air that humans really need to breath.

In order to supply fresh air it is usually assumed that some air exchange must occur (old air out, fresh air in). This process is described as *ventilation* (see Figure 3.1). In times past, this was ordinarily achieved mostly by simply opening the building spaces to some outdoor air supply, permitting the outdoor air to displace some of the interior, semipolluted air. When acceptable for the interior conditions, this is still basically a good method.

However, ventilation of interior spaces in buildings is now mostly achieved with air-handling systems, typically consisting of a piped (ducted) airflow that is pushed into spaces to force some flow of stale air out (called *exhausting* the existing air in the space). In most cases, of course, this flowing air is also "conditioned" in some way—by altering its temperature, its humidity, its odor, etc.

In most buildings there is now some form of forced-air ventilation system—and quite frequently, more than one. In the typical residence, for example, there are probably at least three: one for the general occupied spaces, one for the kitchen, and one for the bathroom. Those for the kitchen and bathroom do only exhausting of the air, while the system for the rest of the house may be a *recirculating* system that literally just moves the air around in the house—"conditioning" it in the process.

The typical air-handling system has the following basic components.

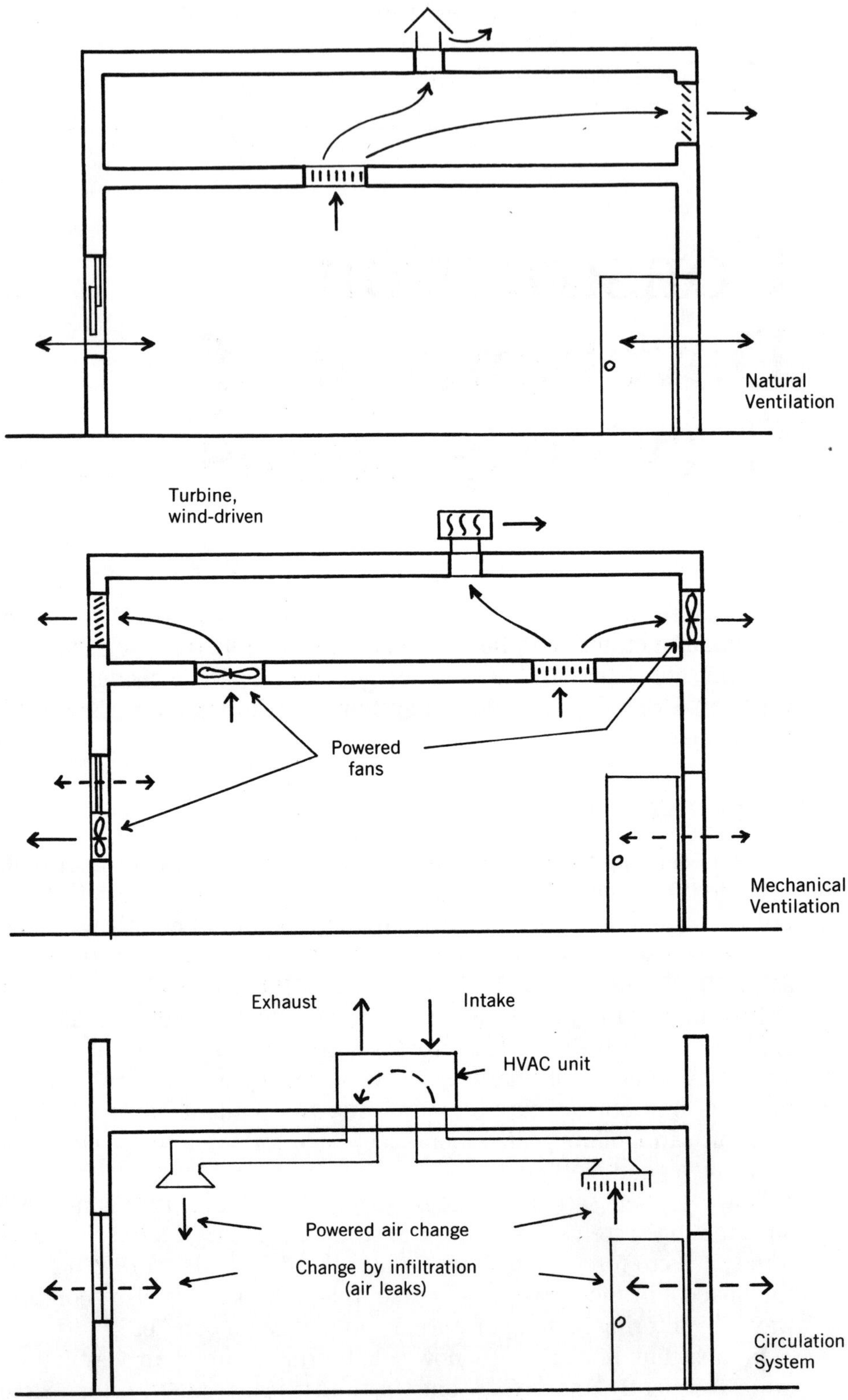

Figure 3.1 Basic methods for achieving building ventilation.

Ducts: These are the "pipes" through which the fluid air is forced, analogous to water in a real piped system. The air enters the ducts, flows through them, and is discharged somewhere. The flow is essentially achieved by differences of pressure in the closed duct system.

Intakes and exhausts: These are points at which openings in the duct system permit the movement of air into or out of the duct system. Fresh outdoor air comes in; stale air is discharged from the building; movement of air into and out of closed rooms is achieved.

Registers and grilles: These are devices used at points of intake and exhaust. They typically perform some functions of control, such as directing the airflow in a particular direction; filtering out particles, bugs, etc.; or restricting the actual volume of the flow.

Mechanical equipment: Fans (actually air pumps) are the principal devices required, although heating, cooling, filtering, humidifying, and other tasks may also be involved.

Controls: Various manual or automatic controls are used. A simple power switch may turn the system on or off. Airflow magnitude (velocity, volume, etc.) may be varied, as well as any of the thermal adjustments or other modifications.

Natural ventilation: This may be deliberately developed, most frequently by simply having windows that can be opened (called *operable* windows. versus *fixed* windows). However, most buildings also leak air in or out; this is referred to as the *infiltration*. The design of the mechanically controlled air-handling system must account for this "natural" ventilation if the whole air situation is to be controlled.

The remaining parts of this section deal with the concerns for these basic components of air-handling systems.

3.1.1 Ducts

Ducts can take up considerable space inside a building, proportional to the air volumes they must accommodate (see Figure 3.2). When the building is divided into many interior spaces, the duct system may be quite extensive, needing to reach virtually every individual space that is occupied. This assumes, of course, that there is a single, central air-handling system for the building.

For small buildings there may well be a single general air-handling system, plus, as appropriate, separate exhaust systems for kitchens, rest rooms, auto parking garages, and so on. The central system may handle both delivery of air (*supply air*) and exhausting of air (*return air*) and function as a recirculation system (round-and-round).

As buildings get larger, it may simply be unfeasible to handle all the building air in a single system. Thus buildings extensive in plan are typically divided horizontally into subunits for air handling and high rises are divided vertically. For some buildings—such as hotels, apartments, and office suites—it may be desirable to use individual systems for each unit.

Duct sizes usually trickle down from the source equipment to the most remote delivery points. Space accommodation must relate to the size of ducts, plus considerations for any insulation and for the duct shape. For piping the fluid air, a round cross

Figure 3.2 Ducts for an HVAC system.

section is best. The farther that any shape gets from round, the worse it is for a smooth air flow (avoiding duct noise and air turbulence).

The flowing air should also preferably not have to turn sharp corners or simply make a great number of detours before being discharged—all in the interest of making the flow easier and reducing the effort required by the flow-inducing equipment.

Ducts are typically quite light in weight, but nevertheless must be supported by the building structure. Attachments to the structure (notably to the underside of roofs and floors) must be facilitated.

Passage of the continuous ducts through the building must be accommodated. This sometimes entails the cutting of openings through some structural elements, such as bearing or shear walls or the webs of very deep beams. This is a situation that requires careful coordination between the various designers so that alternatives can be considered before hard decisions are made for the various building subsystems.

Duct systems must be fire resistive and generally designed and constructed so as not to become passages for flames, smoke, or toxic gases during a fire. This also applies to the penetrations of the construction that are provided for the passage of the ducts, most notably through walls and through floors in multistory buildings.

3.1.2 Air Intake and Exhaust

There is a need to bring quantities of fresh air into buildings and to exhaust stale air from occupied spaces, as well as from rest rooms, parking garages, and so on. This is the basic task of ventilation and it must occur on an ongoing basis, as long as people occupy closed spaces in the building.

To achieve ventilation it is necessary to have a fresh air intake point (or several ones). This should preferably occur at an advantageous point for the obtaining of the fresh air. Locating of such intake points may not mesh well with desired locations for the equipment serving the air-handling system.

Similarly, exhausts must be located at points where the exhausted air will not cause problems for building occupants, pedestrians, neighbors, or the locations of air intake points. In crowded site situations, with buildings very close together, these problems can be of considerable magnitude.

Actually, much fresh air intake occurs from the leaking (infiltrating) of outdoor air into the less than perfectly sealed building. In warm climates, where weather sealing may not occur as completely as in very cold climates, this fresh air source is usually quite significant. With modern construction, however, it is usually necessary to provide deliberately for a specific quantity of fresh air intake to the circulating air system.

Air intake and exhaust points are literally holes into the building volume and must be constructed as such. They may occur simply as screened holes in the building or site surface, or they may be achieved with some constructed element that is connected to the building interior by ducting. Details for these must obviously be worked out in the design of the building's general exterior.

Equipment associated with the air-handling system is frequently located on or near the rooftop, which generally simplifies both intake and exhaust (see Figure 3.3). Still, the separate processes must be effectively achieved without any significant feedback of the exhausted air into the intake.

Figure 3.3 Rooftop HVAC equipment; simplifies exhaust and fresh air intake.

3.1.3 Registers and Grilles

The terminal devices for air discharge in the building (registers) and the generally perforated covers for intake and exhaust points (grilles) must be incorporated into the building design. Registers are usually more numerous, which, when combined with their need for strategic location, makes them highly visible.

Both registers and grilles occur in various forms, and it is sometimes possible to combine standard products with other elements in such situations as integrated ceiling systems (see Figure 3.4). Designers of building interiors must become familiar with the various forms of products as well as with their general range of possibilities for ordered locations.

Both registers and grilles may be developed as special, custom-built elements, although this is generally more easily done with grilles. Registers usually have some special features, such as adjustable openings or directional fins, so their design is a somewhat more technical process. Grilles are usually just screens, often basically just decorative covers for the gaping hole they front.

3.1.4 Equipment

Air handling that consists entirely of the removal of exhaust air from spaces usually requires only fans. These may operate on either a push or pull basis, which relates to their location with respect to the space being exhausted. A critical concern for fan location may be the levels of noise and vibration connected with the use of large fans.

Air for occupied spaces is almost always conditioned in various ways, so the fans are only part of a group of equipment that deal with the air. The equipment may in some cases be collected at a single location, or even actually combined in some form of package unit. Units are usually developed for some specific application and loca-

Figure 3.4 Ceiling registers for an HVAC system.

tion, such as on rooftops, in the exterior wall of the conditioned space, or in a central location serving several spaces.

Equipment can also be separated, however, with different elements located in response to their specific operation, maintenance, and general means of service to the overall HVAC system. There are many different forms of basic HVAC systems, and the type of equipment and its functions vary considerably. For example, heating may be achieved by a furnace and boiler in a basement location, which feeds hot water by piping to coils inside a fan and duct system.

Considerations for equipment that relate to construction issues include the following:

Weight and size of elements that must be supported and housed within the building.

Location of equipment that connects to ducts, piping, and wiring and must be accessible for maintenance and replacement. Ducting especially must run through the building between the equipment and the spaces served.

Noise generated by equipment and within ducts must be considered for its intrusive effects in the building. Noise suppression with enhancements of the construction may be required.

Vibrations caused by heavy equipment, most notably large fans and cooling elements, may cause disturbances in the building, requiring special isolation of individual elements within the construction.

Large groups of equipment housed within the building may have to be isolated from the rest of the building interior in general, for noise, vibrations, excess heat, and so on.

Access to equipment for maintenance may require special consideration when the equipment is essentially built into the construction.

Installation of large equipment may require special consideration when it must be made at some late stage of the construction. This may entail the provision of removable units of floors, roofs, or exterior walls when the equipment is inside the building.

3.1.5 Controls

Controls of many kinds are associated with air-handling systems and with HVAC systems in general. For large installations many controls are automated in some way, with switching controlled by sensors or clocks. The simpler systems may involve some manual control, but complex systems are usually controlled basically only by technically trained people.

Control systems usually involve electrical wiring or tubing for pressure systems. These must be incorporated into the construction in most cases, similar to other wiring, piping, and tubing systems.

Occasionally, the opening or closing of ducts, registers, or other elements of an air-handling system may be manually performed. When the elements being controlled are remote (ceiling or high wall locations), this may entail operating mechanisms that must also be built into the construction.

In general, control systems are no more bothersome than other devices, such as

(a)

Figure 3.5 Devices for ventilation: (a) Louvered vents. (b) Wind driven fans.

power outlets or light switches. They must be installed, however, and their locations are usually somewhat limited by the need for their operation. When they contain sensing devices their locations may be even more limited for proper functioning.

3.1.6 Natural Ventilation

In mild climates natural ventilation offers a major savings in the operating costs of forced ventilation systems. Natural ventilation may be simply provided by use of operable windows. However, special elements or systems can also be incorporated into buildings to make more specifically planned uses of natural ventilation (see Figure 3.5).

A major problem occurs when natural ventilation is combined with forced-air systems. Fan-operated systems work on fluid flow principles and operate most effectively within closed spaces, where pressure differences and air velocities manage the nature of the air movements. Open a window, and the intended air movement paths may be diverted or short-circuited. The fans may end up blowing the conditioned air out the windows instead of into the interior spaces.

The management or successful development of natural ventilation is mostly a planning problem, not much related to the building construction.

3.2 PIPED SYSTEMS

Piped systems include those commonly developed for water supply, hot- and chilled-water service, fire sprinklers, gas, and water or water-borne waste drainage. Piping itself may be achieved with elements of iron, steel, copper, brass, plastic, or fired clay.

Various piping systems must be incorporated into the building's interior construction, usually out of sight in interstitial spaces within the construction (hollow walls, suspended ceilings, etc.). The planning of the piping and its installation details must be developed in careful coordination with the designs for the structure and the general construction. A critical concern in many cases is the ability of the elements of the structure to be punctured for passage of piping, as well as for wiring and ducts (see Figure 3.6).

Piping in the building is connected to some fixture, whose location and installation details must also be developed. Fixtures may be surface-mounted in the construction, but their full development involves the necessary connections within the construction, which may be out of site but quite extensive.

The following discussions develop the concerns for various common systems that require the use of piping.

3.2.1 Water Supply

Water is typically supplied to buildings in the form of what is described as *potable water,* which is essentially water that is fit to drink. Much of this water is not really required to be potable, as it is used for flushing toilets, washing, sprinkling fires, irrigating the landscape, and other purposes not critical in health regards. However, for simplicity, and to avoid the possibility of people misusing the available water source, the single supply is most common.

Water for use in a building may come directly from elevated storage tanks, on the site, on the roof, or on upper floors within the building. However, water supply more commonly enters the building through underground supply piping connected to a

Figure 3.6 Installation of piping within a stud wall system.

water main off the site. Installation of this supply must be coordinated with other site construction and with the construction of the building foundations, basement walls, and so on.

Within the piping system the water is pressurized for delivery at outlets. The pressurization calls for piping that is quite strong and joints that are very secure.

Some special considerations that relate to construction issues are the following:

The piping must be protected from damage, both during construction and during any subsequent modifications of the construction. This is most critical for piping of copper or plastic, but it can affect any piping when major reconstruction is performed. Being buried in the construction makes this a situation that is hard to deal with, but some provisions in the construction detailing may be advisable.

Movements of the building due to thermal expansion, vibrations of equipment, deflections of spanning structures, or lateral drifts due to wind or earthquakes may cause failures of joints in piping systems. Attachments and supports of the piping system must accommodate any significant movements of this nature.

Cold pipes may "sweat" due to condensation of water from the moist air inside a building. This can result in water problems inside the construction, and some protection may be required. Piping for actual chilled water must be insulated, both for energy saving and the sweating problem. However, ordinary supply water is usually ground temperature as it enters the building and stays at some significantly lower temperature than that of the building interior as it flows into the building, so the problem can occur with ordinary water supply piping.

Piping for hot water or steam services must also be insulated, both for energy savings and to protect building occupants wherever the piping is exposed. This piping will also experience considerable expansion and contraction, and its support and attachments in the construction must be designed to accommodate this

For development of piping within the building construction, it is necessary to be aware of the dimensions of elements that require spatial accommodation. The full design of the piping systems will reveal the actual sizes of piping at specific locations. However, some additional considerations that should be borne in mind are the following:

Long runs of piping must be jointed, and the joint elements are typically considerably larger in diameter than the pipe.

If piping is insulated, the full outside dimension of the insulation must be considered; this is especially critical for provision of holes through the structure or containment of piping in thin walls.

Even though pressurized, horizontal piping runs must often be sloped for drainage, making their position within the construction variable. This is especially critical for long horizontal piping runs that must be contained within horizontal spaces in suspended ceilings or must penetrate rows of structural members.

3.2.2 Waste Systems

Piping is commonly provided for two forms of waste. The first comes primarily from toilets and is called *sanitary waste.* The second is sometimes described as *gray water,* and comes from sinks, showers, bathtubs, dishwashers, washing machines, floor drains, and so on—basically, from anywhere but toilets.

In most buildings, these two forms of waste are combined in a single sewer system that carries the wastes from the building. Drainage of this system is by gravity flow, so horizontal runs must be adequately sloped. The slope must continue within the building and out to the sewer main. For buildings that are large in horizontal plan dimension or situated on large sites, this can present a problem.

A third form of waste is that consisting of runoff precipitation from building roofs, balconies, terraces, and paved site surfaces. This is called *storm water waste* and is usually discharged to a separate storm water sewer system or open channel system.

Sewers tend to clog up and need regular cleaning for maintenance or occasional emergency cleaning. Cleanout points are built into the system and need to be accessible.

Drains suck air into the system and develop gases within the system, both of which must be vented to open air by a separate piping system. Individual vents must be provided for every drain point and must run reasonably vertically to the discharge point—typically through the building roof. Traps are used immediately below the drains to keep the foul air and gases from entering the building.

Drainage piping is quite large, increasing as the number of drains that are served increases, to some maximum size at the exit point from the building. As with other piping, the joints are larger than the pipes and must be considered in providing space within the construction.

Large drain piping and the connections that go with it are hard to squeeze into thin walls and must often be placed in larger spaces (called *pipe chases*). These may be clustered with other vertical risers for water or other piping, or even in a partitioned space with vertical wiring or ducts. Grouping of these elements in plan, together with the fixtures or equipment they connect to, is generally a good practice.

Stacking of water-using and drained elements above each other in multistory buildings is usually a good practice, the simplification of construction for vertical drains and vents being a compelling reason.

Swishing sounds in the drain piping can produce unpleasant noise, all the more these days with increasing use of plastic piping and thin wall construction with light materials. Arranging of interior spaces in the building to avoid proximity of quiet spaces to concentrations of piping is the best solution, although sound insulation and acoustic transmission interruptions may be incorporated into the construction in extreme cases. Real acoustic separation is risky business, however, and the truly smart thing is to simply avoid placing sensitive spaces near major plumbing.

3.2.3 Gas Supply

Natural gas from utility mains is piped into many buildings in a manner similar to the water supply: underground to the building base and up through the construction to delivery points. This is also a pressurized system and is highly sensitive to leaks, which in this case can be quite dangerous.

Protection of piping is critical, and the same isolation or other means must be used to avoid deformations of the rigid piping system with possible fracturing of joints. There are many code regulations for the installation of this piping that are intended to promote safety. Any means for protection of the piping from damage during construction or remodeling work will increase safety.

Gas piping is usually relatively modest in size, except for that which feeds large furnaces. It is generally less of a problem for installation than other piping, lacking the problems of sweating of water pipes, venting of sewers, and need for slope in horizontal runs.

3.2.4 Fire System Supply

Fire-fighting systems that use piped water include overhead sprinklers, hose cabinets, and standpipes. Hose cabinets are built into walls—typically in corridors. They generally require piped water that is supplied in a manner similar to other, ordinary water-using fixtures in the building.

Overhead sprinkler systems require extensive piping systems that are supported from the underside of the structure above. This piping must be drained for maintenance, so it can use up considerable overhead space in large areas.

Standpipes are intended to provide water for fire fighters in a manner similar to the familiar curbside hydrants. For low buildings that are large in plan area, they are usually brought to the roof level outside the building. For multistory buildings, they are built into fire-resistant construction inside the building. Codes provide detailed requirements for the systems themselves and for their location and installation details in a building.

Fire protection systems in general are discussed more fully in Sec. 3.6.

3.2.5 HVAC Piping

Various forms of piping are commonly used with HVAC systems. The types of piping required depend on the types of HVAC systems and the size and general nature of the building. Some common uses include the following:

Piping of gas or fuel oil to heating units that are fuel burning

Piping of supply water to boilers for steam or hot-water generation

Connection of boilers with storage tanks and fixtures in steam and hot-water systems

Piping of chilled water to cooling units

Refrigeration systems use piping for the refrigerant and in large systems use water as a heat-absorbing element as well as in chilled form. Systems for cooling typically have several units that may be separated, involving connections with piping.

Piping for steam, hot water, and chilled water must be insulated, adding to dimensional considerations for the construction. Thermal expansion will also be considerable for long piping runs and must be considered in developing supports and attachments to the building structure.

When air systems are used for heating and cooling, it is common to pipe both hot and chilled water to fan coil units within the airflow. This may be done for primary heating or cooling at the central fan system, but in large buildings it also occurs at zone reconditioners, which are used to slightly modify the generally distributed air for a locally conditioned space in the building.

The use of hot and chilled water permits fan coils to be alternately used for both heating and cooling, as demand requires. This requires a four-pipe system to maintain a continuous piping loop to circulate the water, which involves considerable piping to remote units.

The most common rooftop solar heat collectors use water in circulating piping systems to store the heat in large tanks. The same piping system or a second one may then circulate the warm water for heating purposes in the building, for use in bathing, or to raise the temperature of the water in a swimming pool. This is discussed more fully in Sec. 3.9.1.

3.2.6 Miscellaneous Piping

Depending on building and user needs, piping may be used for various additional purposes. Some of these are as follows.

Hot-water service: Most buildings have a general hot-water service, so that sinks, showers, bathtubs, and other fixtures may be supplied by two separate piped water supplies, one for cold water (direct from the main service) and one for hot water (from a water heater). However, for minor usage, hot water can also be supplied by heating units at each fixture.

Special liquid or gas piping: Hospitals, laboratories, or special manufacturing facilities may pipe a variety of other materials through special piping systems. Salt water, gray water, compressed air, oxygen, or just about anything that can be piped may be distributed in this manner. Extensive safety requirements must be followed for these systems to protect users from confusing them with ordinary water piping in the building.

Pneumatic tubes: These systems use a sustained negative pressure (partial vacuum) within a thick tube system to force a capsule through the system. They were used extensively in department stores and offices in times past, but are still used for bank drive-up teller facilities and other places where the system speeds or generally aids some transfer of items. The smooth passage of the capsules requires large radiuses of turns in the tubes, which must be planned for in the installation. The systems should also be exposed as much as possible to facilitate the extraction of jammed capsules on occasion.

3.3 PLUMBING FIXTURES AND EQUIPMENT

Water-using fixtures and drainage fixtures involve the use of piping systems, but they also typically have special installation requirements themselves. This section considers some of the common plumbing fixtures and equipment to which piping is connected in the building.

3.3.1 Toilets

Fixtures for deposit of human wastes include the ordinary toilet bowl (water closet), urinals (for men) and bidets (for women). All of these use water for flushing purpose, hardly needing potable water, but typically getting it in single water supply systems. In seacoast situations, however, where fresh water is in short supply, a separate piping of salt water may be used for these fixtures.

With fresh water in increasingly short supply in many locations, the use of salt water (sea water), gray water from sinks, showers, or other sources in the building, or even systems that do not use water for flushing may become more common.

Toilet fixtures must be securely attached to supports and served by drain systems as well as water supply. Toilet bowls may be floor mounted or wall mounted (see Figure 3.7). Floor mounting involves considerable penetration below the finished floor level to develop the drainage attachments. Wall mounting requires first of all a sturdy wall, but also involves some penetration into the wall space for the attachment of piping behind the fixture.

Bidets are typically wall mounted, although floor mounting is also possible. Urinals may also be either floor or wall mounted, although some local building codes control the options with these fixtures.

The noise associated with the flushing operation of these fixtures is often a source for concern in buildings. Locating of large rest rooms with multiple fixtures must be carefully planned so as not to create an ongoing annoyance or unpleasant situations for some interior spaces. Special fixtures or simply care in the design of the piping systems and their installation can alleviate the noise problems, but distance is the best design tool here.

In addition to the mechanical problems of these fixtures, there are needs for concern of maintenance of the spaces in which they are installed. Floor and wall finish materials must be selected for resistance to water, staining, and corrosive effects.

3.3.2 Sinks

Sinks range from the simple lavatory sink in bathrooms to special sinks for various purposes: kitchens, bars, janitors' closets, and various industrial or other unique uses.

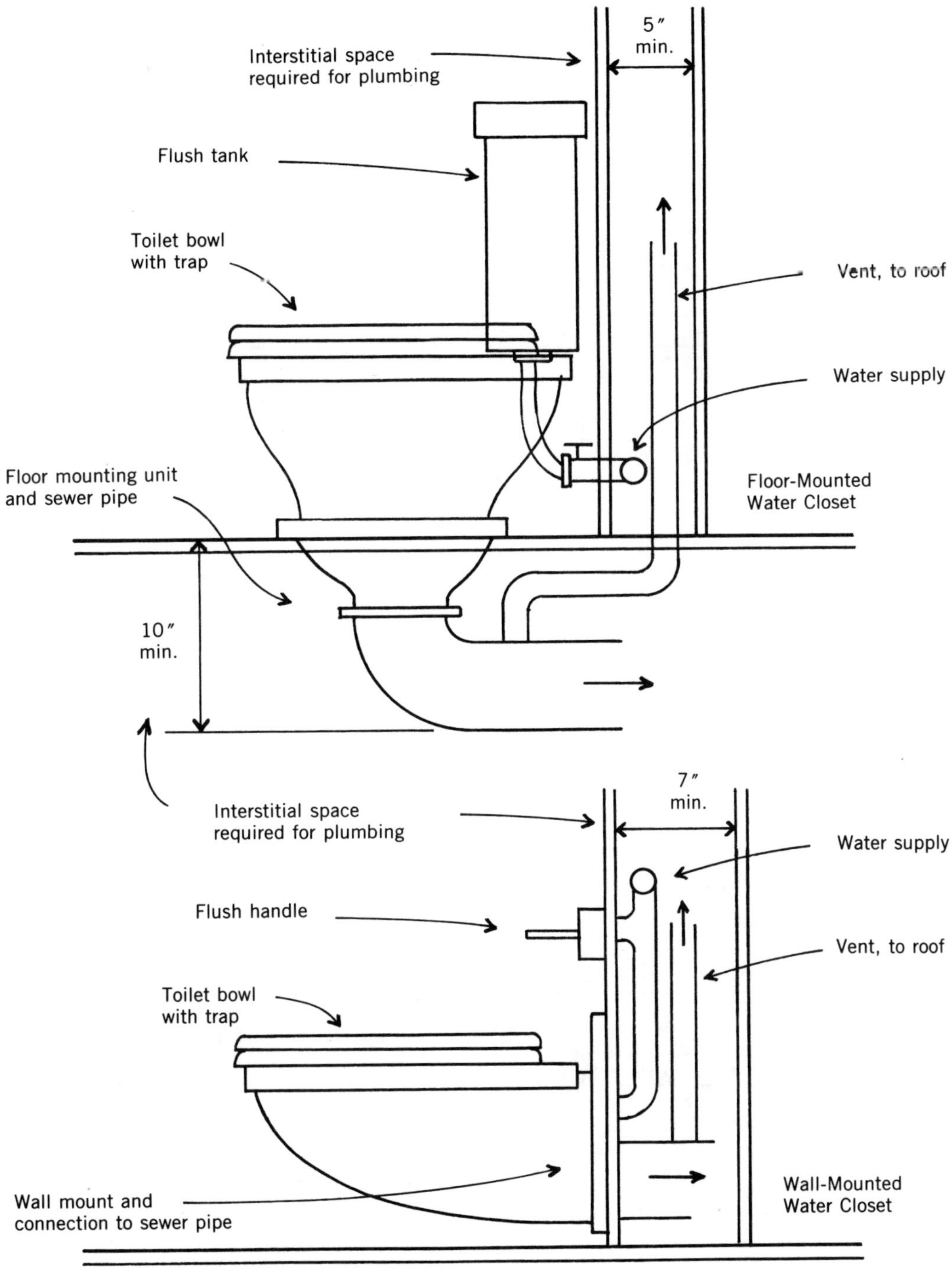

Figure 3.7 Structural and plumbing considerations for toilet installations.

The ordinary sink has both cold and hot water service and a drain that feeds the common sanitary sewer in the building.

Sinks may be independently floor or wall mounted, but are also frequently built into some form of cabinet or counter top. Although not quite as structurally demanding as wall-mounted toilets, a sink that is wall mounted does require a relatively sturdy wall for support.

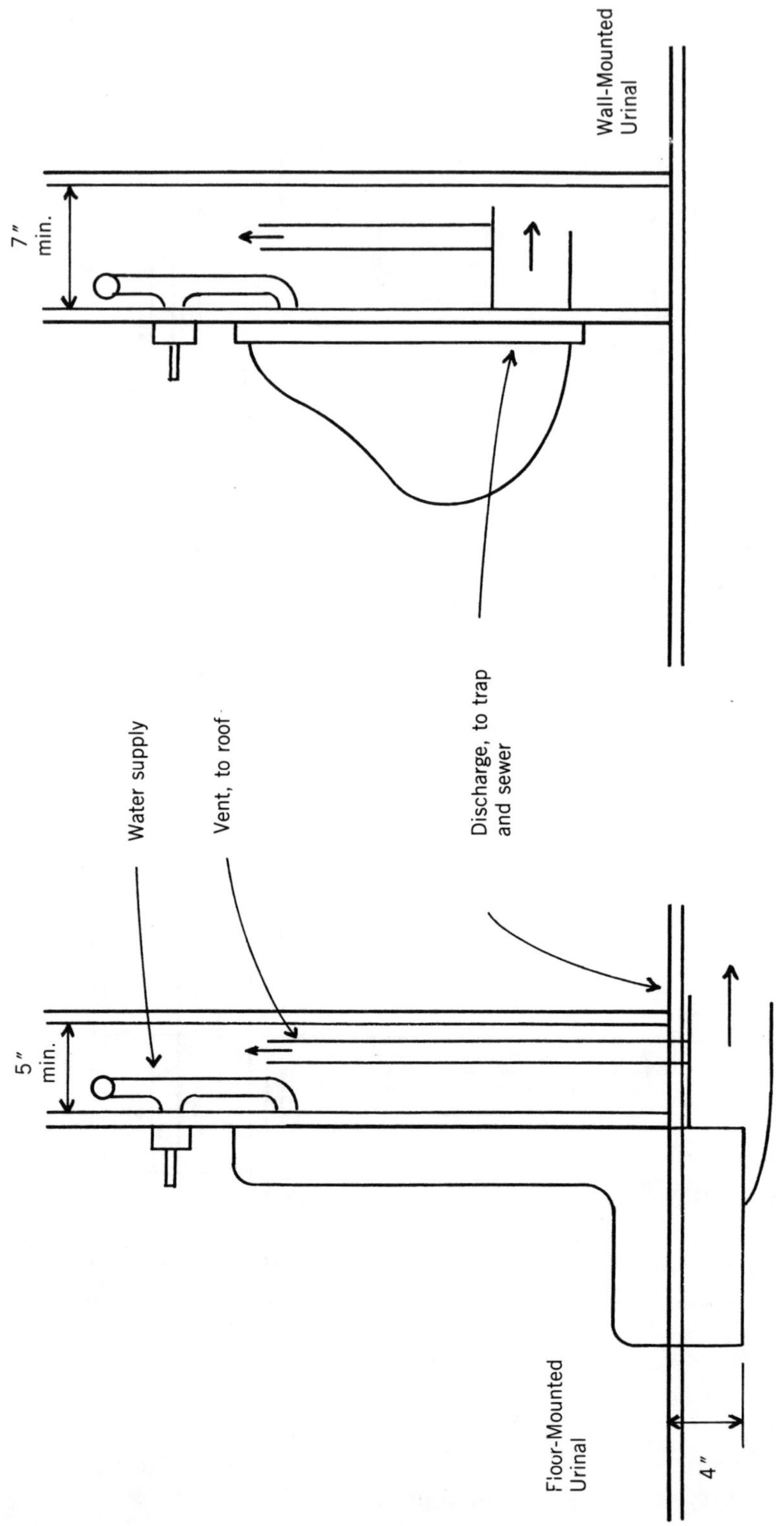

Figure 3.7 *(Continued)*

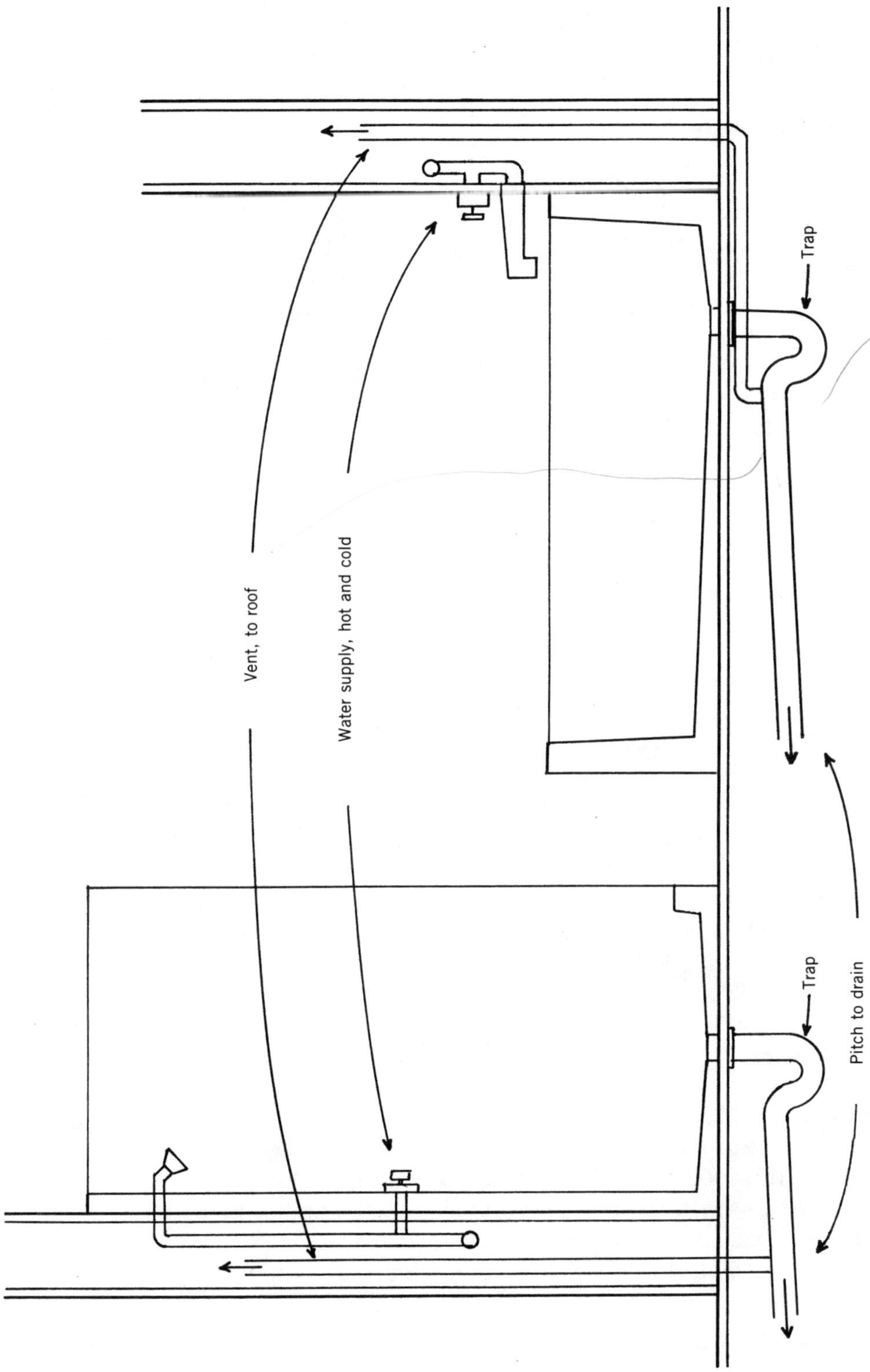

Figure 3.8 Details for installation of tubs and showers.

Kitchen sinks usually have garbage disposal units in the drainage channel and also frequently serve the drainage discharge from dishwashers. The electrical wiring, water service piping, and drainage piping must be facilitated in the general construction. This is commonly done with some ease when the fixtures are all built into a cabinet and counter top, but need some study when sink units are freestanding.

3.3.3 Tubs and Showers

Bathtubs, hot tubs, Jacuzzis, and showers all require considerable accommodation in the construction. Ordinary concerns are for the cold- and hot-water service, drains, and structural supports. A bathtub or hot tub full of water is very heavy, and when placed on upper floors, constitutes a considerable load.

Drains for showers and tubs usually require some depth below the floor line for their installation (see Figure 3.8). If the drain piping has a long horizontal run, the total depth increases. Floor structures with relatively shallow depth may well not provide this extent of interstitial space.

In multistory buildings dropped ceilings are frequently used in bath and rest rooms to accommodate overhead exhaust fans as well as plumbing from fixtures above. (See details for the bathrooms in building 4 in Chapter 5).

3.3.4 Drains

Sinks, bathtubs, showers, and toilets all use drains. However, drainage is sometimes necessary as a single function, not associated with some fixture. Flat roofs, rooftop terraces, parking lots, and some general floor areas may be drained to a collection point at which some form of drain is provided.

So-called *area drains* take various forms, depending on the nature of the area drained. For roofs, a common form is that shown in Figure 3.9, with a screening device generally formed to reduce clogging in a location that often does not receive maintenance cleaning and tends to become cluttered with debris. The upward protruding shape of the drain screen is not a problem for the rooftop, which is not supposed to facilitate traffic.

For floors or other locations where traffic does exist, the drain screen must be reasonably smooth. The size of the drain and the type of traffic to be borne will determine the general structural nature of the drain and its cover. Figure 3.10 shows a type of drain sometimes used, consisting of a trench with linear drain cover units. Single-point drains take forms similar to that shown for the roof, except that a flat cover is used.

Considerable spatial accommodation is usually required for area drains to provide for the large drains as well as the large piping connected to them. For roofs this space must be provided within the roof support structure. For floors and other locations the appropriate space must also be provided. When built into paving slabs, it should be assured that any foundations or other construction beneath the paving will not present problems for installation of the drains and piping.

3.4 WIRED SYSTEMS

There are three major systems in most buildings that involve installation and operation with wiring. These are the basic electrical power distribution system, the lighting systems, and the communication systems (phones, intercoms, etc.).

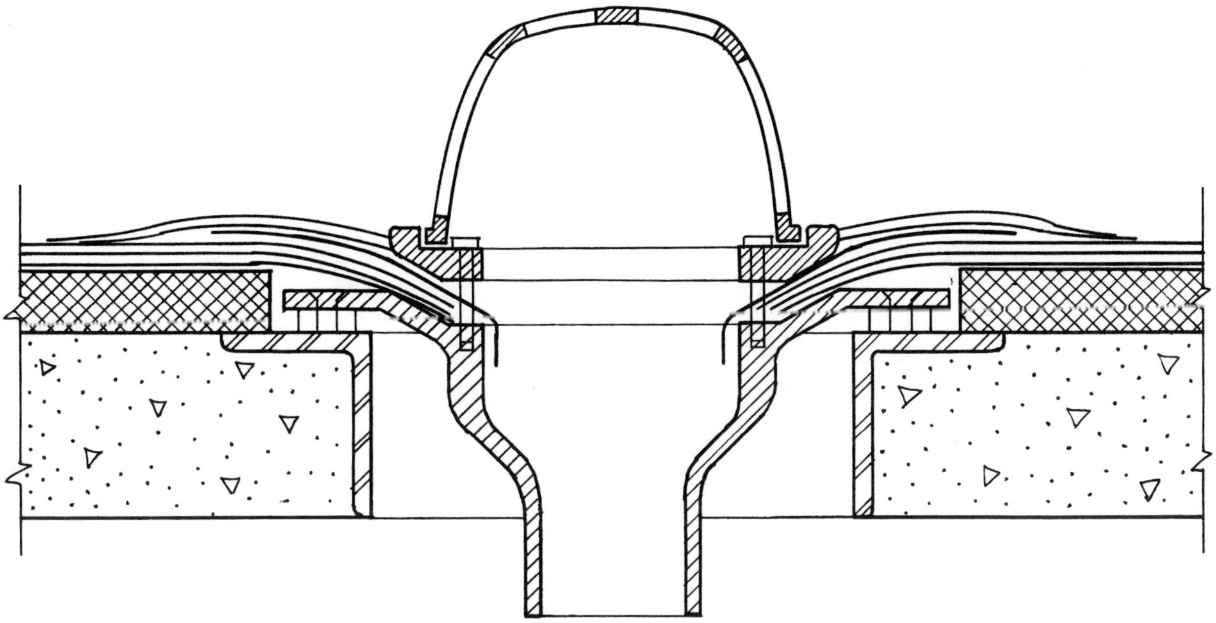

Figure 3.9 Details for an area drain on a roof; piped to an interior drainage system.

Wiring for all of these systems is typically installed in some semipermanent fashion during the initial construction of the building. However, as user needs and available technology change rapidly, it is common to need to upgrade or modify the systems on an ongoing basis; thus the wiring must lend itself to this eventuality.

The wiring and its installation represents an entity for various construction concerns. In addition, the elements served by the wiring (power outlets, light fixtures, phone jacks, etc.) must be considered, and in some cases installed in the construction. Also, these services are usually connected to larger systems outside the building, so incoming wiring must be dealt with as well.

In addition to the three wired systems mentioned previously, there are often many other uses of wiring in the building. The following parts of this section consider various uses of wiring and the construction issues to be dealt with for the wired systems.

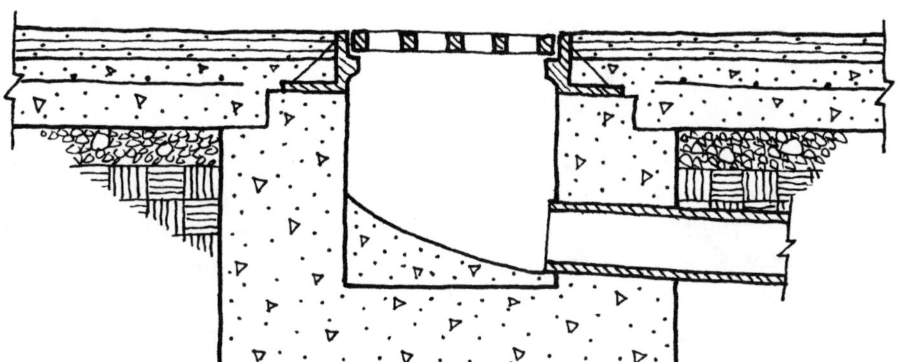

Figure 3.10 Detail of a trench drain in a driveway.

3.4.1 Electrical Power

Almost every building uses electrical power for many purposes, lighting and operation of equipment being principal and common uses. For remote sites or for emergency power, the source may be on the site. However, the more common situation involves bringing the power to the site in some form of transmission from a distribution network.

A principal concern in development of construction details related to electricity is safety. The two aspects of safety of greatest concern are the possibilities for fire and for electrical shock to people. Also of concern is the possibility of damage to equipment due to excessive electrical current, a major economic concern when the equipment is very expensive.

Principally because of safety concerns, but also for insuring against equipment losses, extensive controls are maintained over the materials and construction installations for electrical wiring and all the parts involved in the electrical systems. The codes used for this are national ones, and every building designer should have some understanding of their principal provisions relating to construction.

Some primary concerns for construction with power supply systems include the following.

Protection of wiring: Wiring must be protected from damage, moisture intrusion, and anything that may cause an electrical short. This is basically achieved with parts of the wiring system (sheathing, conduits, boxes, etc.), but also relates somewhat to its positioning and general installation in the construction. Attachment and support of the more permanent elements of the electrical system should be provided by the more permanent elements of the building construction (not suspended ceilings, not nonstructural partitions, etc.).

Accessibility of wiring: As mentioned previously, the wiring systems should be accessible for ongoing modifications. This is not the exception; it is the rule. The wiring systems themselves are designed and built for this, but the construction may or may not be as accommodating. Tearing up of permanent construction *will* be done if necessary, so it is best that it not be necessary.

Housing of equipment: Equipment required varies considerably, depending on the size of the electrical service and complexity of the systems. Once the size and general nature of a building project are determined, some early estimates can be made regarding the necessary elements of the electrical system and their potential locations. For code requirements or simple functional use, some locations will be fixed: inside or outside, rooftop or underground, and so on. Locations of major equipment generally also represent locations of major amounts of wiring, which must be incorporated into the construction.

Wiring for specific pieces of equipment or fixed-in-place lighting fixtures must be provided in forms and with precise locations responding to the planned items. Power outlets for the plugging in of movable equipment or appliances must also be provided. Codes now specify minimum requirements for general power outlets as well as locations of light switches and other use-related details of the systems. Both the outlets or switches themselves as well as the wiring to them may sometimes be difficult to incorporate into the construction, especially with those forms of wall construction not having easy-to-use interstitial spaces.

3.4.2 Lighting: Power and Control

Building lighting is typically achieved with some combination of daylight and electrical illumination. Issues regarding use of sunlight are treated in Sec. 3.9. This section treats the problems of the electrically powered lighting systems.

For wiring, the lighting systems generally break down into two distinct systems (see Figure 3.11). The first is *hard-wired;* that is, it is done with fixtures that are attached or built into the construction, with appropriate wiring permanently installed and specifically supplied to each fixture. The second system is movable and highly variable and uses the general power outlets in the building.

Fixed (hard-wired) lighting may be minimal in some buildings, with the user left to provide movable lighting elements for most purposes. More commonly, a minimum number of fixtures are built in to provide general illumination and some specific task lighting (for bathrooms, kitchens, front doors, stairs, etc.), with movable lights anticipated for special tasks.

In some buildings the major lighting may all be built into the construction, providing completely, at least in theory, for all the anticipated tasks. However, the power outlets are always there and can still be used at the occupants' discretion.

The basic forms of lighting systems and their construction implications are described in the following subsection. For wiring alone, a principal concern is the control of the lighting, as well as its basic power supply. The most common form of

(a)

Figure 3.11 Wiring for lighting: (a) Hard wired—installed in the construction. (b) Plugged into the general power outlet system.

(b)

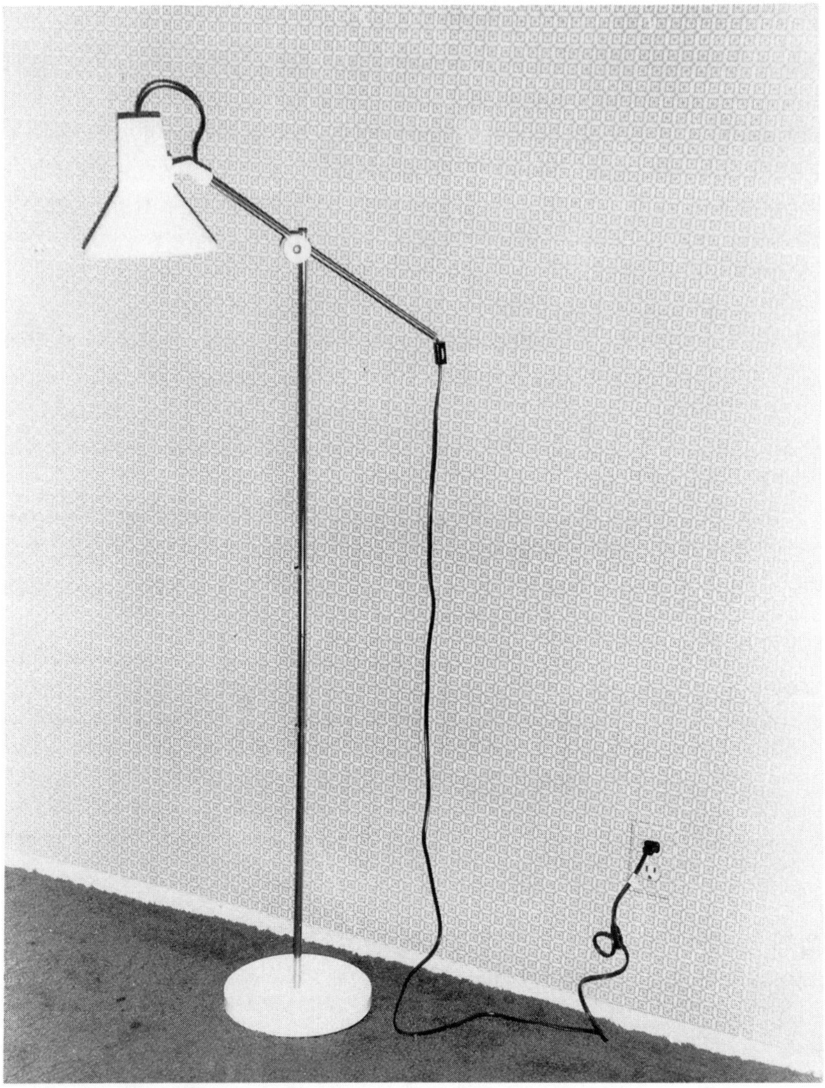

Figure 3.11 *(Continued)*

control is by manual switching, with switches for each unit of lighting that needs controlling. Switches may be easy for anyone to use or may be secured in some way to prevent use by other than authorized persons.

Wiring for powering the lighting and for switching must generally be built into the construction in a permanent fashion. Switches are usually in walls, and the fixed lighting may be in walls or in ceilings. Stringing of wiring through the walls and ceilings thus becomes a construction planning and detailing consideration. As in other cases, the forms of hollow construction that provide easy-to-use interstitial spaces will accommodate this with the most ease. Solid constructions of masonry and concrete offer some greater challenges on occasion.

Wiring, as well as the switches and lighting fixtures, may be surface mounted rather than built into the construction. For interiors with less concern for quality of finishes, this may offer a simplification for the eventualities of modifications, as well as some easier initial installation. The preceding will be the more significant when the basic wall construction does not so easily accommodate built-in wiring.

3.4.3 Lighting Systems

Individual lighting fixtures, both permanently installed and movable, abound in great variety. For purposes of construction considerations alone, there are some basic types of fixtures and systems. Since movable fixtures relate only to the power outlets, we consider only those elements that must be accommodated directly by the construction.

Figure 3.12 shows the basic forms of individual fixtures, as well as some special constructions for the development of lighting. For fixtures that are surface mounted, the construction accommodation consists primarily of the wiring and the connecting devices provided for attachment and electrical hookup of the fixtures. Exceptionally large fixtures, such as large pendant-style chandeliers, may require considerable structural support, but small fixtures are often structurally supported by the connecting boxes installed with the wiring.

Required accommodation in the construction becomes greater when fixtures are partly or fully recessed into the wall or ceiling surfaces. This becomes largely a matter

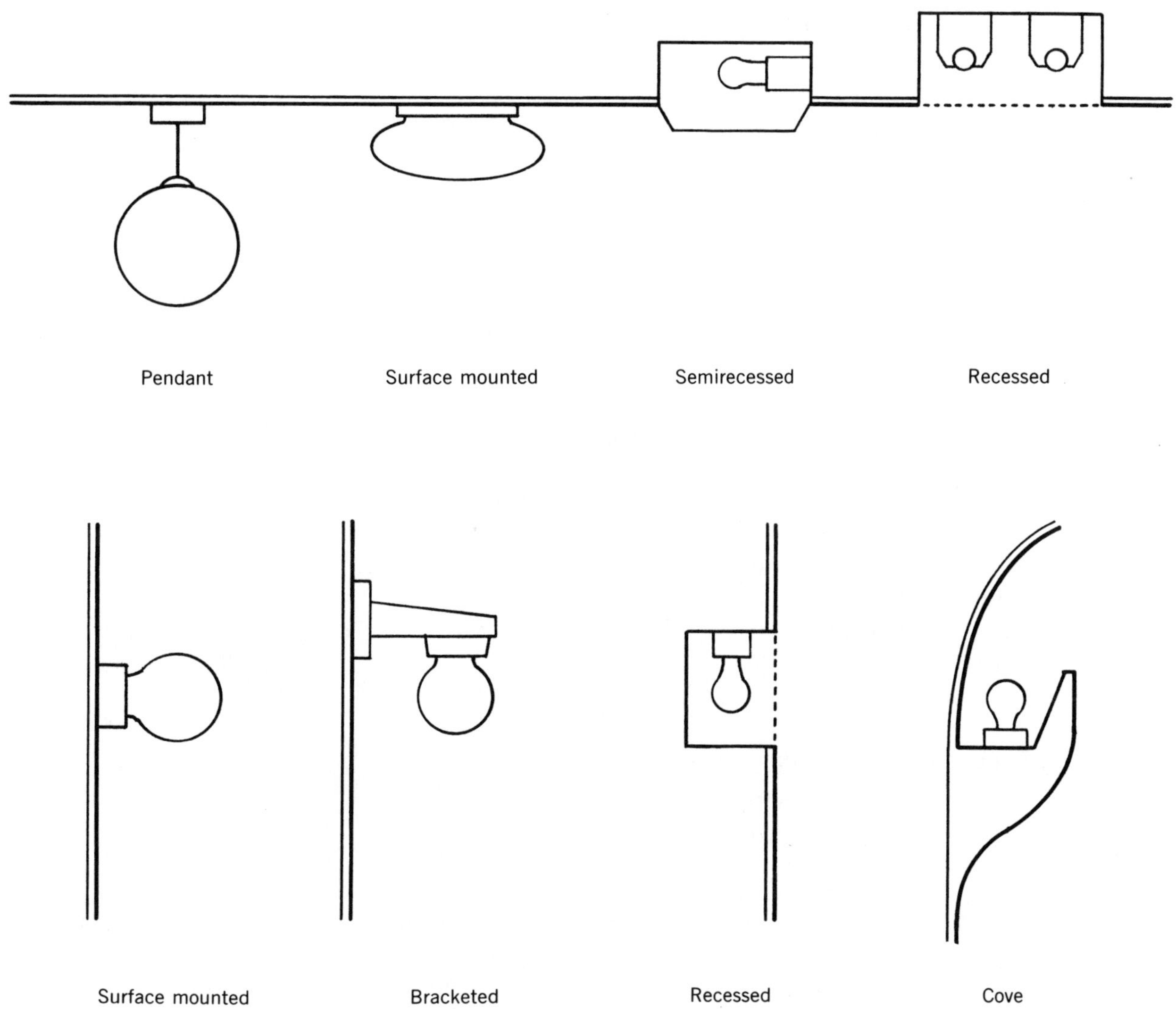

Figure 3.12 Basic forms of lighting fixtures.

of the spatial dimensions required, although the heat generated by the lamps may also require separations for fire safety. All of this is easiest with suspended ceiling systems having generous interstitial space. In other situations, the space may be hard fought for or truly difficult to provide in the form of construction of the wall or of ceilings that are basically the underside of a spanning structural system.

Exterior lighting and sometimes special interior lighting may require special structures, such as tall poles or cantilevered brackets, for installation. These structures may be constituted as independent, freestanding structures with their own requirements for foundations and general construction.

Many integrated ceiling systems are developed to incorporate modular lighting units, many of which are produced to fit into such systems. Priority systems are also available with fully integrated ceiling structures, finish surfacing, lighting units, and HVAC units. Products for these systems have certain specific requirements and modules, but also permit a wide range of variations.

3.4.4 Signal and Communication Systems

In addition to the lighting and basic electrical power systems, there are typically several other wired systems in the buildings. Some common ones include the following:

Telephone, cable TV, computer network, intercom, and other communication systems

HVAC sensing and control, security, fire sensing and alarm, doorbell, and other signal systems

Many of these systems are built into the construction, which involves the wiring installation as well as the development of the various devices that service the systems.

Some systems utilize the basic electrical power system of the building, either directly or through transformers that reduce the voltage. Other systems, such as telephones, utilize their own separate power sources. In any event, the circuits within the systems are usually independent and the wiring of each system is a complete and separate arrangement.

For systems with lower voltages, the safety issues of fire and electrical shock are much reduced, and the safety requirements for the wiring and general installation are much less severe than those for the basic electrical power system. Connections must be adequately developed and wiring protected to some degree, but the construction details for the installations are generally somewhat easier to achieve.

For various reasons, the individual wiring systems must be kept separate. However, the general spaces provided within the construction can be shared in many cases—with other wiring as well as with piping and other buried elements. The traffic can get jammed in tight spaces, but sharing of interstitial space is a common situation in most buildings.

A special construction provided for frequently modified wiring systems is the *elevated floor*. This consists of a working floor surface developed with modular units on top of short posts that sit on the general spanning floor structure. Installation of such a system is shown for building 7 in Chapter 5. Easy removal and replacement of the modular floor units permits ongoing modifications of the wiring with relative ease.

The elevated floor was originally developed for use in "computer rooms," but as the computer has steadily invaded the whole work space, it has become feasible to use it for entire floors, particularly in office buildings and research facilities.

3.5 SOUND CONTROL

Management of sound in buildings has steadily increased as a major design issue in recent times. Some reasons for this are the following:

Increasing use of powered appliances and equipment, many of which generate some sound in their operation

Replacement of old heavy forms of construction with new and generally lighter ones, steadily reducing the inherent sound isolation and barrier effects of ordinary building construction

Proliferation of sound-producing devices (radios, TVs, cassette and CD players, etc.) and extensive use of them by many people

General trends toward development of tighter planned spaces (small lots for houses, more apartments, more people in urban settings), which result in people being crowded closer and closer together

This trend results in the aggravation of several potential, basic sound problems in buildings, the principal and common ones being the following:

Noisy conditions experienced as discomfort by individuals who desire or require some degree of sound privacy. Noise is essentially any *unwanted sound,* whether it comes from screeching tires or a symphony orchestra.

Difficulty of obtaining secured sound privacy, for the protection of private communications and so on.

Difficulty in obtaining some specific desired sound conditions in a particular space (lecture hall, library reading room, recording hall, etc.).

Difficulty of isolating or blocking some specific unwanted sound sources, such as large fans, flushing toilets, slamming doors, or low-flying aircraft.

The full management of sound can in most cases be only partly achieved by proper design of the building construction. Building planning that strives to place major noise-producing items as far as possible from sensitive sound areas is usually more effective than modification of the construction. Having cooperative, sympathetic fellow building occupants or neighbors is even better.

Assuming that every other possible measure has been taken and sound is still a problem, there do in many cases remain some things that can be done to improve the sound character of the construction. The following discussions present some of these considerations.

3.5.1 Sound Performance of Construction

Just about every imaginable assemblage for roof, floor, or wall construction, using currently available materials and products, has been tested and rated for various basic properties—including responses to sound. Primary ratings used for design include the following.

Transmission resistance: This is the ability of the whole construction to reduce the level of an airborne sound or a surface impact sound as it passes through the construction, from one side to the other.

Surface absorption or reflectivity: This is the effect of the construction on sounds that bounce off of its surface; basically a property of individual surfaces, it is sometimes affected by the whole structural character of the construction.

Transmission resistance is important to establishing sound privacy of individual spaces. Surface absorption generally affects the character of sound in an enclosed space. Transmission may be significantly altered by changes in the construction. Interior sound conditions tend to be more importantly affected by room shapes and sizes and the points of origins of sounds.

Every space, at a given time, usually has some enduring sound level (called the *ambient sound level*), sustained by running equipment, ongoing conversation, and so on. This ambient sound may make hearing difficult on occasion, but also frequently tends to mask much unwanted sound. To some extent, therefore, what is essentially undesired sound (noise, that is) may actually serve a useful purpose in screening other, more annoying sounds or in helping to assure sound privacy.

Attempts to modify sound conditions sometimes backfire; that is, the end result is worse in some ways. What is a marginally objectionable ambient sound level may be modified by alterations or enhancements of the construction, with the result being a different, and possibly even more objectionable, sound condition. This is a design game to be played by persons knowledgeable about sound and its consequences, as well as the mechanical properties of materials and constructions.

Some very common objectionable conditions can be avoided by a combination of good planning and some attention to the details of the construction. These issues are discussed in conjunction with the development of the nine case study buildings in Chapter 5.

3.5.2 Enhancement of Construction for Sound Performance

Every form of construction has some inherent characteristics with regard to sound. These attributes may be utilized by selection of basic construction elements for their improved sound-related properties. While this may be possible in some cases, where very special sound problems exist, there are often other more compelling reasons for choices of the construction.

In many cases it is possible to improve the properties of the basic construction for sound responses. This may be achieved by adding elements, such as insulation or vibration isolators, or by some modification of the basic elements. Modifications may consist of changes in thickness, density, surface form, stiffness, or other properties.

For improvement in the reduction of airborne sound transmission, there are generally three major areas of improvement that may be achieved. These are described as follows.

Density

Generally, heavier construction is more resistive to airborn sound transmission. Opting for a heavier form of wall or floor construction (masonry, concrete, etc.) will generally achieve less transmission and greater sound privacy or isolation of noises. Adding concrete fill on top of wood decks—as is now commonly done for apartments, motels, and offices—improves various properties (less bouncy floors, better fire separation, less live load deflection), among which are more resistance to both airborne and surface impact sound transmissions.

Vibration Interception

Sound travels through a solid wall partly by causing the wall to vibrate. Airborne and surface impact sounds delivered to one surface are transmitted through the wall and vibrate the opposite surface and thus the air in the space on the other side of the wall. Part of this vibration is lost in traveling through the materials of the wall, but some devices may be used to further lose some of the vibration as it travels (see Figure 3.13). Some form of insulation added to void (air) spaces in the wall may help, but the most effective devices are those that work as breaks in the vibration path (shock absorbers, vibration isolators, etc.). The basic objective is to make one wall surface independent of the other, as much as possible.

Vibration is also reduced by travel through more dense material or thicker elements; thus efforts to increase the wall mass tend also to produce basic reductions in the transmitted levels of vibration.

Sealing of the Barrier

For the most part, sound travels from one space to another by direct air passage—through leaks in a barrier, or by a flanking action around the barrier (out the door, into the hall, back in a door; out a window, back in another window; into an air register, down the duct, out another register). This is probably the number one item to pursue in attempts to improve sound separation for privacy or noise isolation.

Sealing of roofs and floors is more easily achieved than sealing of walls. There is generally no such thing as a nonstructural roof or floor, for example, so the construction tends to be more structurally continuous and rigid. Nonstructural walls are often not continuous to the underside of the overhead construction if dropped ceilings are used.

All walls tend to pull away from their boundary edges, by shrinkage, temperature expansion and contraction, and the natural settlements and lateral movements of the structure. Walls are routinely pierced for outlets for electricity, telephones, cable TV or computer networks, and for light switches, HVAC registers, thermostats, return air grilles, and various built-in fixtures.

Minor modifications of the design and general improved care in the construction are major means for improving sound responses. This is true for ordinary situations and may be sufficient for a generally enhanced performance. Even for very special problems, where extra elements and special details are provided, attention to the simple problems itemized above will provide extra response enhancement.

In many cases major efforts and investments in extra construction are made to enhance barriers, only to have leaks or flanking actions subvert the whole effectiveness of the work for sound improvement. It is therefore advisable that these basic efforts be made for *all* construction, ordinary or special.

3.5.3 Vibration and Noise Control

Sound, as sensed by humans, is mostly vibrating air. Efforts to control sound vibrations by modification of the construction generally affect the whole character of

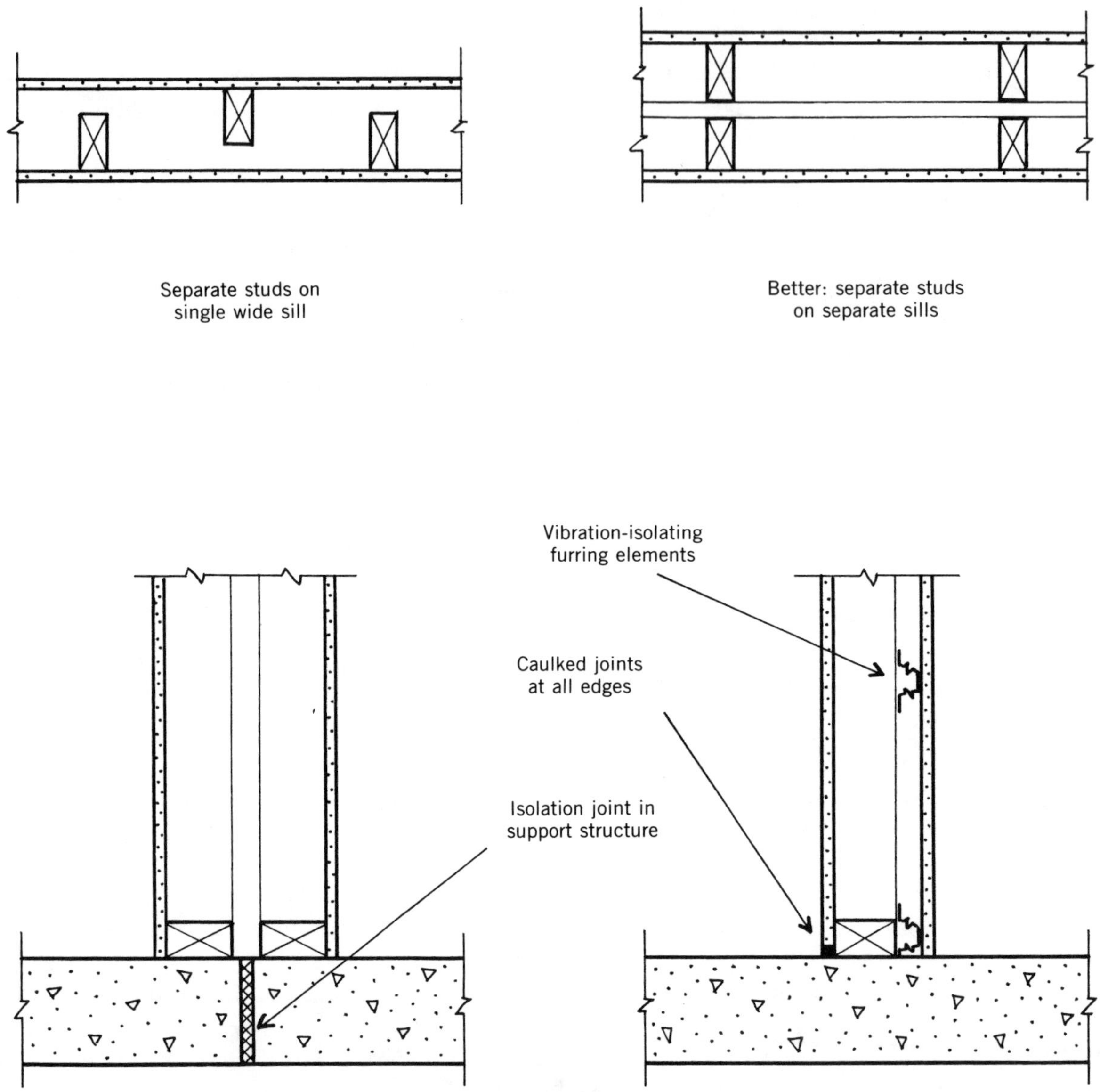

Figure 3.13 Development of wall construction for interruption of sound vibrations.

vibration responses. In buildings, some major sources of vibrations produce both noise (sound, airborne) and direct vibration of the building structure. Both of these effects may be objectionable, and both frequently need attention.

Protection from noise (of a person at some specific location in the building) may be achieved at the noise source (by isolation), along the path of travel of the noisy sound (by barriers or vibration interceptors), or by protecting the hearer (with ambient noise levels, ear plugs, etc.). For really critical levels of unavoidable noise (large air-conditioning compressors, large fans, ducts carrying high-velocity moving air, etc.), all forms of protection may be necessary for really effective achievement of sound privacy.

Direct sensible vibrations (shaking) may be more difficult to control in some situations. These are often transmitted by the building structure, with the structure's own relative stiffness, continuity, and tightness of connections being properties that make the transmission more efficient and serve to carry objectionable levels of vibration great distances through the construction. Structural separation joints—for thermal expansion, seismic control, or structural deformation control—may intercept vibrations, but in general the properties that make the structure work well as a structure tend to also make it work well as a transmitter of vibrations.

Direct vibrations, as well as significant levels of noise, may often be most effectively controlled by dealing with the source. Equipment that produces major vibration and noise is usually rated for the level of its output, which may be used as a design criterion for selection. Location of equipment in the building may be a major tool for isolation, simple distance of separation being a critical issue. Placing of equipment in lower levels of multistory buildings may generally isolate it from a major portion of the buildings' occupied spaces.

This is a complex problem, but a common one for most buildings. Every house, office, and school has problems involving the unwanted intrusion of sounds and vibrations from flushing of toilets, swishing of air in ducts, slamming of doors, ringing of phones, zooming, honking, and screeching of traffic, and so on. In just about every case, *something* could have been done with little effort to improve the construction for response to these situations.

3.5.4 Management of Interior Sound

The management of interior sound conditions in buildings involves all the areas of consideration, including the following:

Good planning to separate annoying sources of noise from areas where noise is most objectionable. Don't put the rumpus (look it up!) room next to the master bedroom.

Selection of basic construction to improve sound conditions—where reasonable alternatives exist and sound is a major concern.

Use of minor modifications to enhance sound performance for critical elements of the construction (partitions between private offices, floors between apartments, hall and exterior walls for lecture rooms, etc.).

Attention to planning and development of construction details to eliminate sound leaks and flanking (run partitions to the underside of the overhead construction; don't place outlets, air registers, or bathroom wall cabinets back-to-back; etc.).

Use of vibration isolators and/or sound-shielding enclosures to isolate noisy elements (furnace fans, air-conditioning compressors, kitchen and bathroom exhaust fans, etc.).

Planning of the shape and dimensions of rooms to avoid bad sound situations (echoes, resonance, focusing, etc.).

Use of room surface finishes to improve sound behavior in the room; generally meaning the use of sound-absorbing materials where possible (carpet, not ceramic tile, etc.).

An elementary knowledge of sound behaviors can help designers to avoid the simple problems that are very common. Really critical situations need extensive study by qualified professional acoustic consultants.

3.5.5 Management of Exterior Sound

Sounds incoming to the building include those generated on and off the building site. Management of the interior/exterior, site-to-building interchange may be subject to some design efforts. These may deal with the locating or enclosure of any noisy exterior equipment, separation of buildings in groups, locating of noisy site activities (children's playgrounds, etc.), or use of site planting or site construction (walls, earth forms) to achieve some separation.

Noises incoming from heavily traveled streets, other buildings, flying aircraft, or other off-site sources cannot usually be affected at the sources, which leaves basically the efforts of some good planning or enhancement of the building's enclosing shell. With an ever-increasing amount of new construction occurring in crowded urban locations, this form of enhancement for buildings is becoming almost a routine situation. Ordinary forms of wall construction and the construction elements used to achieve enclosure are rated for acoustic transmission, and respect is paid to this value in many designs.

In many buildings, windows are a principal source of entry for exterior sounds. Many of the efforts made to achieve good weather seals or thermal response also improve resistance to sound transmission, which helps to reinforce choices justified previously only for HVAC considerations. Use of laminated and double glazing is of particular note is this regard, especially for multistory office buildings where the percentage of glazed wall surface is usually quite high.

Where low-flying air traffic is considerable, the sound barrier effect of roofs becomes important, especially for low-rise buildings where the roof surface may constitute a major portion of the total building enclosure. The whole sandwich of roofing, insulation, structure, interstitial space, and ceiling must be analyzed for this total effect, making for many possible variations to enhance the construction.

3.6 FIRE PROTECTION SYSTEMS

Achievement of fire safety is generally viewed mostly as an effort to make the building itself fire resistive and to facilitate the speedy evacuation of occupants in the event of a fire (see Figure 3.14). These are indeed major design concerns. However, many buildings also contain various forms of fire protection and fire-fighting devices and systems. This section addresses some of the construction implications of these systems.

3.6.1 Detection and Alarm Systems

Most building fires are short lived and have only localized affects. One reason for this is the wide use of various systems for detecting fires and for calling the existence of the fire to the attention of occupants and others (security people, building supervisors, maintenance workers, fire fighters, etc.). Detection may be achieved by building occupants or by automatic detection devices on the basis of heat, smoke, oxygen depletion, and so on.

Detection systems usually require minimal accommodations by the construction, except for wiring and surface mounting of devices. However, devices must be strategically located and their existence acknowledged in the full development of the building interior design.

Alarm systems may be manually activated by building occupants or be automati-

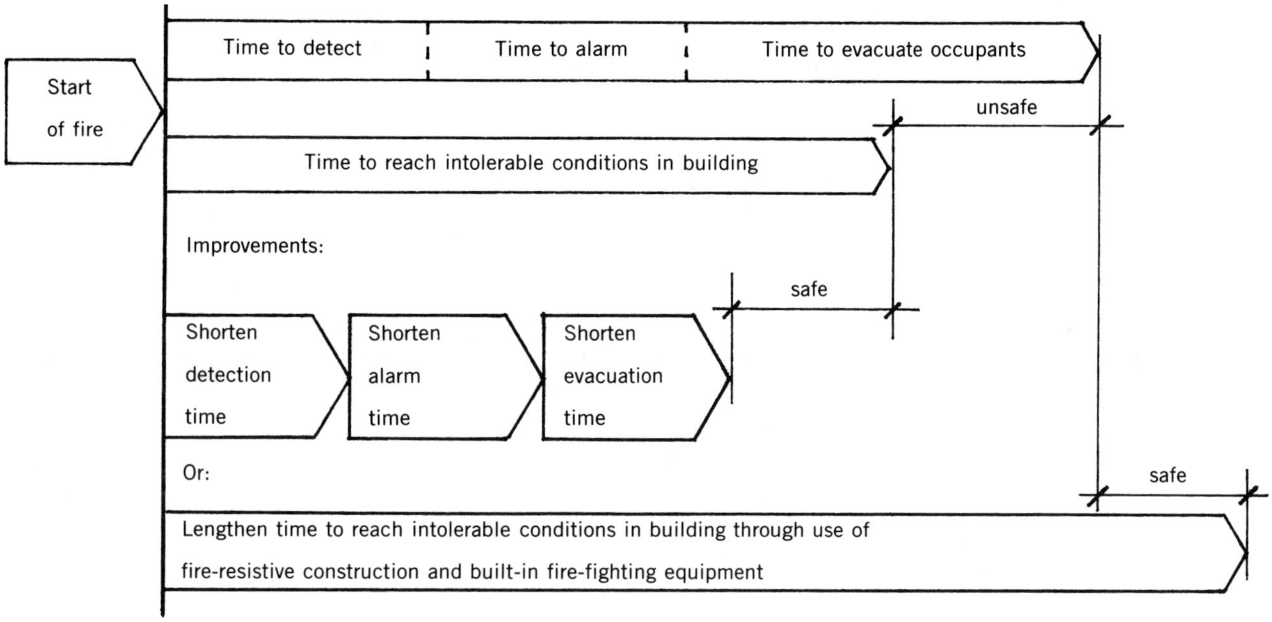

Figure 3.14 Fire safety; viewed as a race against time.

cally activated by detection devices. For noisy building interior conditions, it should be ascertained that the alarms can be effectively detected by occupants in order to achieve their attention-grabbing task.

Design of interior construction should not result in blocking or other reduction in the effectiveness of either detection or alarm systems.

3.6.2 Fire-Activated Elements

Detection devices may be used to activate certain controls or fire-fighting systems. Some common elements developed for this action include the following.

Elevators: The basic intended exit path for multistory buildings uses the stairs, not the elevators. To prevent people from becoming trapped in elevators, detection devices sometimes interject to deactivate the elevators or to make them go to the ground floor level and stay there unless released for use by fire fighters.

Fire doors: These are generally doors that are designed to be closed automatically to prevent the spread of fires. In special situations, some doors may be designed to *open* automatically to assure a free exit path for occupants.

Blocking for ducts: Fire dampers may be placed in some ducts to prevent the ducts from becoming paths for the spread of fires, especially from floor to floor in multistory buildings.

Vents: Most deaths of building occupants during fires result from smoke or toxic fumes. In some situations the use of automatically opened roof or wall vents may draw off the smoke or gases; however, these can also cause a chimney effect to increase the intensity of the fire, so their use must be carefully studied for real safety effectiveness.

3.6.3 Fire-Fighting Systems

Various systems may be used for direct, automatically activated suppression of fires or as general aids to occupants for fire fighters in their efforts to fight the fire. Automatic water sprinklers are a commonly used system of this type, generally activated by thermal-sensing devices. Similar systems may be activated by smoke or other actions or provide special chemicals or other materials more appropriate than water for special fires.

Hose cabinets, usually built into corridor walls, are required for some buildings. Besides being built into the construction and connected to water piping, these must be strategically located, so planning for them is restricted. Occupant-activated emergency alarm devices are also required in some cases, also requiring strategic locating.

Standpipe water supply systems are also required for some buildings, providing points of attachment for hoses for fire fighters at locations less likely to be served by the usual outdoor fire hydrants at ground level. These may be placed on the outside of a building, as shown in Figure 3.15, but where this is not desired, they can sometimes be placed in stairwells or other interior locations.

Requirements for special fire-fighting equipment depend largely on the size of a building and on the number of occupants. They also frequently depend on the location of the building with respect to neighboring buildings, fire zones, street access, and so on.

In some design situations choices for special fire-fighting systems may depend on

Figure 3.15 Water supply system for fire-fighting.

the chosen construction and its effects on fire insurance charges. Risks—actuarial and real—may be reduced by use of detection, alarm, and various special systems, even though they are not required by codes or zoning. Use of certain materials, plan arrangements, or construction details may be possible if some extra precautions are taken in these regards.

Sometimes the particular concerns of owners or users of buildings for extended fire resistance may indicate the use of features not required by regulations. Remember that building code requirements are in general *minimal* in character. If some level of safety beyond a minimal one is desired, owners may feel the investment is worthwhile.

3.7 VERTICAL TRANSPORTATION SYSTEMS

People and various objects must be moved vertically in multiple-level buildings, often requiring the use of special equipment such as elevators, escalators, or special material-moving systems. These typically represent major elements that must be incorporated into the construction, occupying considerable space in a continuous form—often from the foundations to the rooftop. The following discussions consider some issues regarding construction development to facilitate installation and operation of equipment for vertical transportation.

3.7.1 Elevators

Elevators may be used for moving both people and materials. People-movers are described as *passenger elevators*; they may also be used for some moving of furniture and other materials on occasion. Special elevators used primarily for moving materials are called *freight elevators*. These two are the most common forms of building elevators, although many special types are also used.

For powering of the vertical movements, there are two basic types of systems. One employs an overhead, counterweighted suspension system using steel cables and electrically powered motors and sheaves. (See Figure 3.16.) The other basic type uses a large hydraulic piston that is usually in a chamber beneath the elevator cab.

Overhead suspended elevators are capable of relatively rapid travel and quicker operation in general; thus they are sometimes referred to as *high-speed elevators*. They are also capable of achieving considerable total vertical distances of travel, which when combined with their relatively quick operation, makes them the choice for tall buildings.

Hydraulic elevators operate a bit more slowly and are used mostly for only a few stories of lift. The two elevator systems have different space and equipment needs, however, so there are often compelling reasons for using one over the other. One of the requirements for the overhead suspension system is the need for considerable height of the elevator shaft space above the highest point of entry to the cab, often resulting in a tall structure on the roof when the top story is relatively short. Where the spatial intrusion at the roof is undesirable for architectural design, or even sometimes for functional reasons (severe building height limits, for example), conditions may strongly favor the hydraulic elevator—unless, of course, the building is very tall.

Both elevator systems require considerable construction below the level of the lowest entry point to the cab. If the elevator serves the lowest building level, this can mean a considerable depth of excavation and construction below the basement floor.

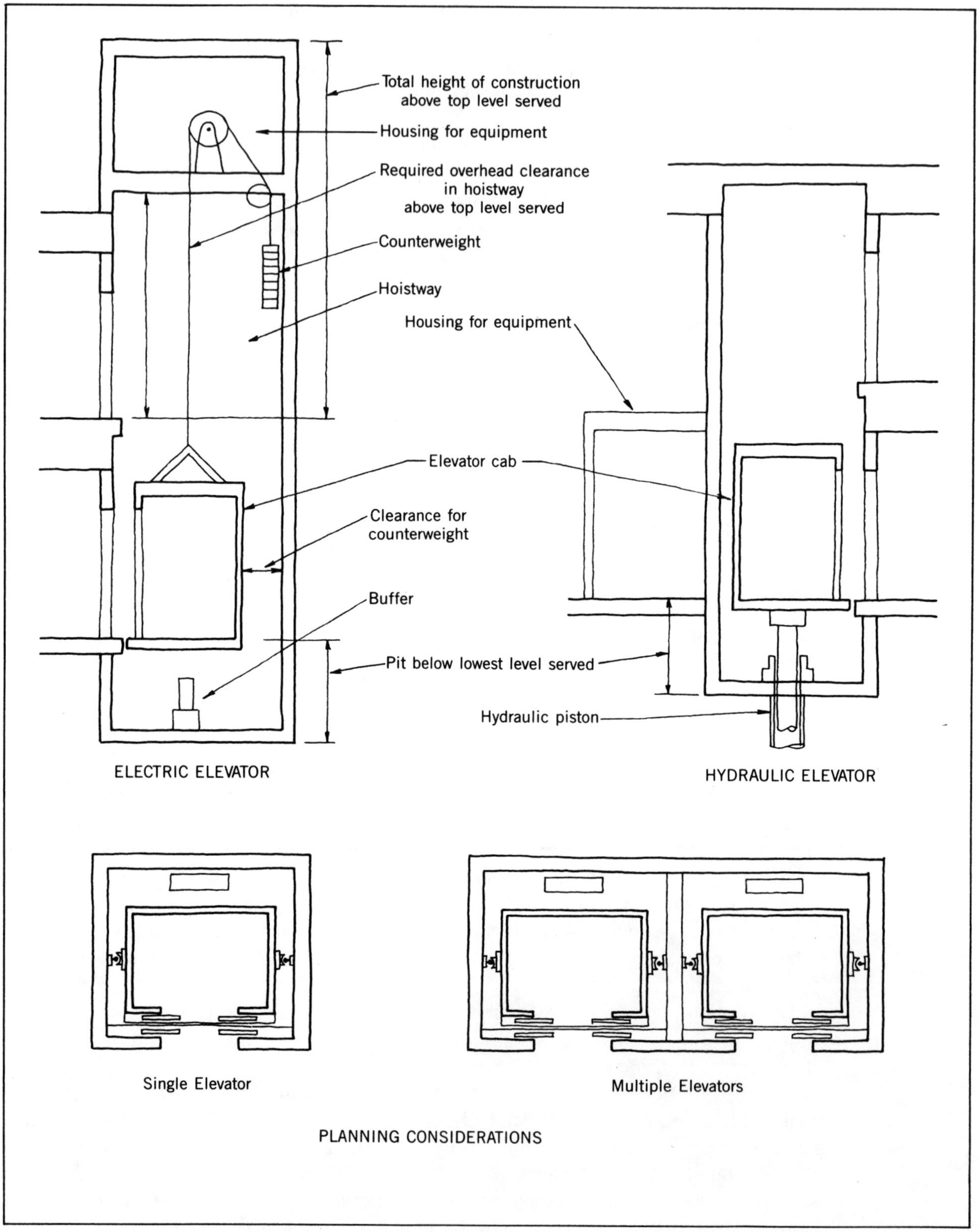

Figure 3.16 Considerations for incorporation of elevators in a building.

Design of the building foundations in the area near the elevators must recognize this situation.

Obviously, the elevator shaft must be vertically aligned, and thus it becomes a constant presence in all levels of the multistory building. When the building is quite tall and supporting columns are close to the elevator shaft walls, any considerable change in column size from top to bottom must be done in a manner that does not intrude on the interior wall face of the elevator shaft. Planning of corridors and other spaces in the vicinity of the elevators must also recognize this problem.

Elevator cabs move along vertical guide rails, and the mechanical operation of the moving cabs, the entry doors, and the operating equipment for moving the cab are all sources for noise and vibrations. The supporting structure and general surrounding construction must handle these effects with a level of protection for the rest of the building in terms of sensible noise and movements during the operation of the elevators.

The dimensional requirements for elevators vary considerable, depending mostly on the cab size and the total number of elevators used in the system. For a small two- or three-story building with a single elevator, the space required may be small. For a very tall building, with a large number of elevators, the space at lower levels may be a major portion of the whole building plan.

3.7.2 Escalators

Moving stairs (escalators) are used in various situations for various reasons. They sometimes offer a relatively uninterrupted path of movement for pedestrians, compared with the delay represented by waiting for an elevator. They are also often used in atriums or other open interior spaces, where their openness to the view may be a consideration.

When not operating (moving), escalators can serve as stationary stairs, so their slopes are generally subject to the same limits that exist for stairs. The plan space required for an escalator thus begins with that required to achieve a stair for the total height of movement. Beyond this, space is required for the return loop of the moving steps and the housing of the operating equipment. (See Figure 3.17.)

As with elevators, the operating equipment and moving-stair system involve noise and vibrations, which must be dealt with by the structure and separating construction if close-by areas are to be protected. As compared to elevators, the operating equipment exists at each level for a multistory escalator system, so the noise source is more generally spread through the building.

3.7.3 Material-Moving Systems

Material-moving systems vary considerable. The ordinary freight elevator basically emulates the passenger elevator in general form, although size, load capacity, door operation, and other details may vary considerably. Other systems may be used for special materials or special means of movement. Very large elevators may be used for moving vehicles or other large items. Smaller systems may be used for the movement of small objects, with no accompanying human passengers.

One of the smallest systems is the pneumatic tube system, which sucks a special capsule into a smooth tube and propels it to some receiving point. This was used extensively in department stores and offices in former times, but is now seeing some recurrence of use as a simple means for rapid collection of materials in some

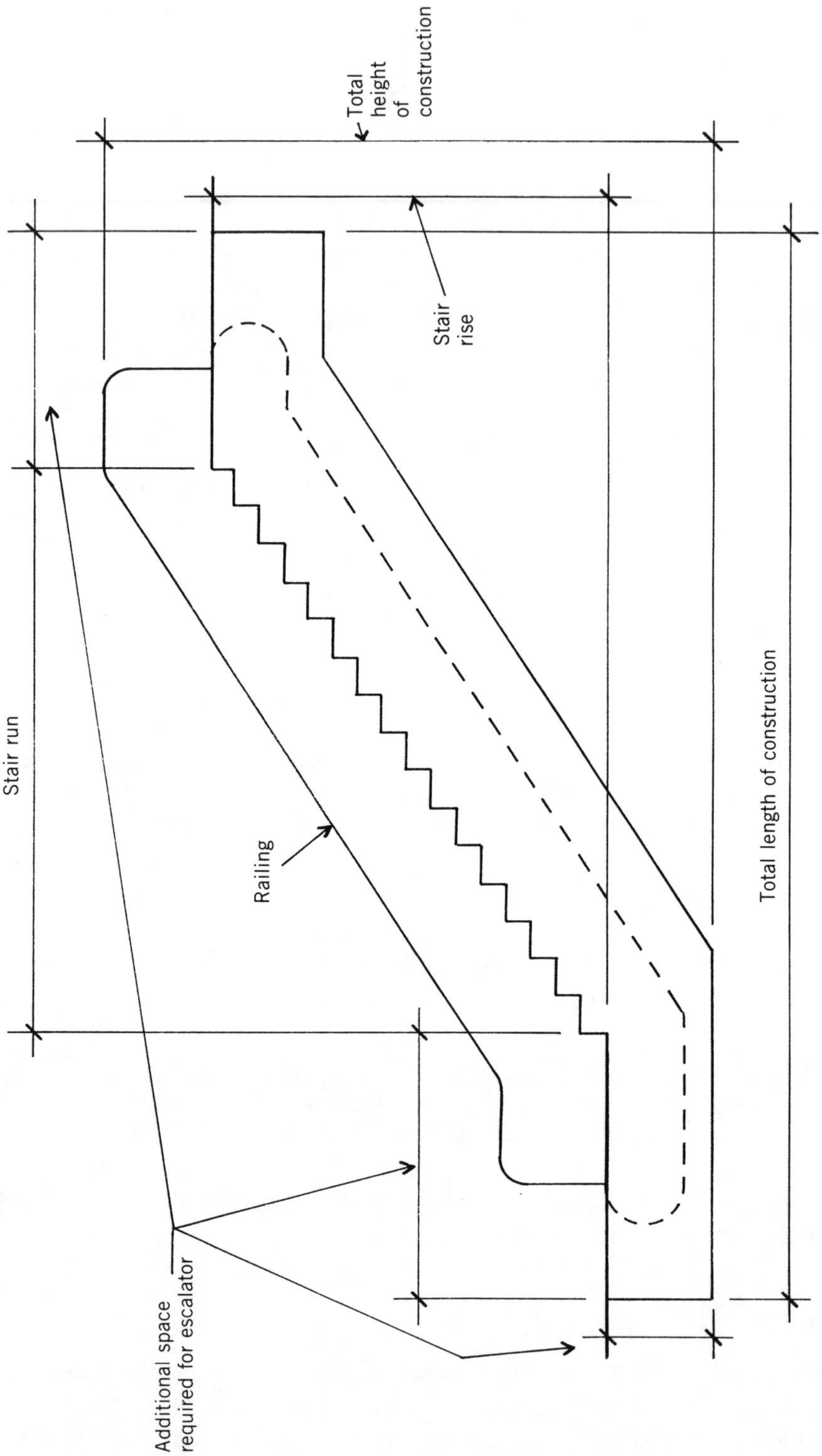

Figure 3.17 Considerations for incorporation of escalators in a building.

operations. Total distance and the number of turns in the path are limited, and the turns need some radius, so installations must be carefully studied and accommodated spatially in the construction, somewhat analogous to piped systems for water drainage.

3.8 THERMAL CONTROL SYSTEMS

Of all the building "services," the oldest and most universal is that of the systems used for control of indoor thermal conditions. Most of the basic systems for achieving this control have long histories in architectural design and building planning and construction. The details of the technology change, but basic functions and needs generally remain the same—the primary one being the need to maintain some bracketed range of thermal conditions for health and comfort inside the building while outdoor temperatures fluctuate over the climate's daily and yearly ranges.

Incorporation of the various elements used to achieve thermal control (generally constituting what we call the HVAC system) typically has a significant impact on the development of building interior forms and many details of the construction. There are also major concerns for the construction itself—principally that of the building's enclosure—with regard to its passive participation in the buffering of the internal thermal condition against severe external conditions. The remaining parts of this section treat some major issues relating to the building construction and its role in thermal control.

3.8.1 Thermal Performance of Construction

Maintaining thermal comfort for building occupants generates various concerns for the enclosure construction, primarily the roof and exterior walls and windows. Some primary considerations are the following:

Resistance to thermal flow, that is, the natural effect of the materials and the form of construction on direct air-to-air flow of heat into or out of the building

Air leaks, resulting in the heat loss or gain represented by the volume of air change, cubic feet per minute flow translating to calories or Btus per hour losses or gains

Miscellaneous effects; such as those of the interior surface on air movements, radiant losses to cold interior surfaces, and so on

Ability of the construction to accommodate elements of the building's HVAC system

Specific climate conditions, the general form of the building HVAC system, and the needs of the occupants will determine the relative importance of various factors and the general need for or effectiveness of various forms of enhancement of the enclosure's basic construction.

Development of any enhancements of the building enclosure begins with an analysis of the properties of the basic construction. If the results of such an analysis produce values for behaviors that are less than satisfactory, then some modifications or additions are considered. Of course, smart designers will anticipate needs and make some considerations of these issues in their initial choices to avoid disruptive changes later. Routine, repetitive use of construction with demonstrated successful performance will give some assurance that additional modifications of a major nature will not be required later.

Routine usage, however, needs to be monitored for the effects of developments in new technology and changes in design criteria as code changes, new industry standards, research, and new design procedures affect the relative value of the "good" old ways of doing things.

A large problem for the designer is that of determining the truly best forms of construction for a specific application. What works well for an office building in Minneapolis cannot be expected to be ideal for a church in San Diego. Even worse, what works for one church in San Diego may not work for another, depending on its location, sun orientation, size, shape, and so on.

Superinsulation of walls may be effective in very cold climates, but where cooling is the major issue, air infiltration and solar heat gain are typically the principal concerns. In fact, adding to the construction may in many cases be less critical than proper choices and correct detailing of basic construction systems and elements.

Manipulations of the materials and the details of the building construction deal essentially with the passive responses of the building to physical actions. Considerations for passive effects must be included in the design of the building's active (mechanically operating) HVAC system. Judgments about the effectiveness of various materials, details, or basic methods of construction should not be made by the architectural designer, detailer, or specifications writer without inputs from the designers of the HVAC system.

Designers of the basic building construction have vital roles in the development of passive response as a thermal control issue. As conservation of energy has become increasingly important, this aspect of construction design has become better supported with data and design criteria and procedures. Common forms of construction have tabulated properties for various effects: thermal flow, insulative value, heat storage capacity (heat inertia effect), and resistance to air leaks and moisture migration. Designers need to be aware of these properties to the same extent that they are aware of fire ratings, deadweight, and other significant properties of common construction options.

Moisture Migration and Condensation

Airborn moisture passes to some degree through many materials, other than truly water-repelling ones. This presents various potential problems that must be considered in the choices for materials and their arrangements in the construction.

A major problem develops when moisture passes through or into construction that is between two spaces of considerably different temperature. Warm, relatively moist air moving to a colder location may reach a point where some of the water held in the air will condense due to lowering of the temperature. If this occurs inside construction, the accumulated water can cause damage. Over time, the damage will be progressive and eventually result in some serious condition.

One technique for reducing this effect is to place a generally water-repelling, continuous surface element in the construction to impede the flow of the airborne moisture. Such an element is called a moisture barrier, and depending on the construction, the degree of the problem, and climate conditions, various techniques may be used.

In some situations, ordinary elements of the construction can be used as moisture barriers, an example being metal roof decks. If these have their joints well sealed, they are very resistive to the moisture movement through them. In other situations, it may be necessary to introduce special elements to achieve the moisture barrier function. Items used for this purpose include aluminum foil, plastic film, impregnated felts, and coated paper. Depending on the details of the construction, the barrier element may be added as a separate piece, or it may be used to surface some other part of the construction, such as wall sheathing or an insulation unit.

A critical concern in using moisture barriers is to have them situated in the construction so as to impede airflow before a major temperature change occurs. If warm, moist air comes up against a cold moisture barrier, condensation will occur on the face of the barrier. If this is inside the construction, the water condensed can create a problem.

Condensation can also occur inside hollow elements of the building construction, such as stud-formed voids, cavities in masonry, or the insides of hollow mullions in curtain wall systems. In these situations it is sometimes possible to simply let the moisture condense and provide for its drainage with the necessary shaping and sloping for flow control and use of flashing and weep holes to control its passage out of the construction.

This is a really major issue of concern in development of the construction for the building enclosure, and there is a great deal of information available for study and design development.

3.8.2 Thermal Enhancement of Construction

As stated previously, all basic forms of construction for roofs, windows, and exterior walls have rated properties regarding thermal performance. One design procedure, therefore, is to shop between various options for basic construction systems or elements with desired properties. This may indeed be done, but there are also various ways to enhance or improve the properties of basic types of elements. Means for achieving this usually consist of one or more of the following:

Minor alteration of the form or details of the basic construction, such as increasing of the void space in a stud wall by using wider studs or increasing of the thickness of individual layers in a sandwich-type, built-up unit

Substitution of different materials for the usual component elements of a system, such as using a more insulative panel material instead of plywood for wall sheathing or roof decking or using some less thermally conductive material in place of metal for framing of a curtain wall to reduce the thermal bridge effect

Adding of elements to the basic construction, such as separate insulation materials or vapor barriers

The need for and relative effectiveness of any individual change must be considered. Doing all of the above modifications to a single wall will have a cumulative affect, but the cost and other factors for each must be considered in terms of the benefit achieved. Widening the studs thickens the walls and reduces the usable interior building volume for the same exterior dimensions; thus the whole unit cost of the construction per usable space is altered, not just the cost of the wall itself.

3.8.3 Heating and Cooling Equipment

Incorporation of equipment for thermal control is a major concern for planning and in some cases for detailing of the building and its construction. HVAC equipment is bulky, heavy, makes noise, vibrates, and needs to be accessible for service. It also needs attachment to wiring for electrical power and signal controls and to piping for water and gas or oil fuel lines.

Location of equipment must be reasonably strategic for its operation in the HVAC system, which sometimes creates problems of a structural nature or serious situations for noise and vibration control. A large fan or an air-conditioning chiller or condenser near a space where noise or vibration is highly undesirable poses a problem for careful consideration. The preferred first choice is to achieve all the distance of separation that is is possible. After that the possible options for improvement all have feasible limits of effectiveness and often considerable expense.

In addition to the equipment itself, there are usually heightened noise problems associated with the piping or ducts in the vicinity of the equipment. These will usually be at their largest size and maximum flow rates, presenting both dimensional and maximum noise problems. Major efforts may be made to isolate the effects of the equipment itself, only to lose the battle with screaming ducts and gurgling pipes.

3.8.4 Miscellaneous Systems

Depending on the type of basic HVAC system used and the size, shape, and use of the building, there may be many special subsystems that exist in conjunction with the basic heating, cooling, and air management systems. Some common ones that need consideration in building planning and detailing of the construction are the following.

Control Systems

In the age of automation, HVAC systems typically use forms of control that are quite sophisticated—even for small buildings with simple purposes. Control can be as simple as flipping a single electrical switch (on or off; that's it!). However, most installations employ monitoring devices (thermostats, etc.) that are typically connected to automatic switching devices. The ultimate extension of this is the so-called *smart building* that not only automatically monitors existing conditions, but anticipates events such as the impending arrival of the building occupants for work or the predictable swings of daily temperature.

Control is often much more than just switching of equipment on or off. Fan speeds may be adjusted; flow rates in ducts may be adjusted by movable dampers; localized adjustment of air temperature in ducts may be achieved with zone reconditioners; discharge registers may have fins permitting a range of opening from fully open to fully closed.

Many controls are operated with electric signaling devices using low-voltage current. Installation of wiring is thus less restricted than with that for the higher-voltage power lines. Nevertheless, the installation must be anticipated, and the need for strategic location of monitoring devices and any hand-adjustable control units must be recognized.

Solar Energy Input

Systems using the sun as a direct source are more fully discussed in Sec. 3.9. These systems may be the primary sources for energy, but they are also frequently either supplementary to or supplemented by other powered systems. In any event, the installation requirements for solar systems are often quite elaborate; and, of course, strategic location of the solar-collecting devices is essential.

Thermal Storage Elements

Heat can be stored and some form of deposit and withdrawal system may be established. Systems using hot water or steam have always used this principle, with storage tanks for ballasting of the system. Many more uses are now made of this concept, especially the use of passive storage to ride out the daily temperature swings. Thus the heat of the sun during midday is used to warm some mass, from which heat can be extracted during cool nighttime periods. Or, a mass is exposed to cooling at night and then used to absorb heat during the day. This is generally called the *thermal inertia* principle.

A practical source for a thermal inertial mass in some cases is the building construction. For this purpose, mass translates simply to weight, and thus the heavier the construction, the greater the potential for heat storage. This obviously favors concrete or masonry over light, hollow frame construction, and the issue must be dealth with in selection of construction options where this performance is a design goal.

Waste Disposal

Several forms of waste are generated by HVAC systems, common ones being the following.

Exhaust air: This is usually "waste air" extracted from occupied spaces, the refreshing of which is what ventilation is all about. It may also contain undesirable elements, such as odors from toilets or fumes from parking garages. The latter type of air must usually be completely dumped out of the building. Less fouled air, however, may be treated or slightly refreshed with cleaner air (usually outdoor air) and returned for repeated use to occupied spaces.

Fumes, vapors, etc.: Burning of fossil fuels or other activities may produce fouled air or other gases that must be vented from the building.

Wastewater: Condensate from dehumidifiers or other forms of fouled water may be treated as waste and dumped into the building's general wastewater disposal system.

Incorporation of elements for waste disposal are sometimes major concerns for planning of the construction. Wastewater lines must flow by gravity and be vented; hot chimneys must be insulated or isolated from combustible construction elements and be extended through the roof; fouled air must be dumped out where it will not disturb the neighbors or short-circuit back through fresh-air intakes. Planning for the basic elements of the operating HVAC systems must also anticipate the need for these various disposal components.

Fuel or Energy Storage

In former times anticipated need for generating heat meant the need to store vast quantities of wood or coal—still a problem where these are used. A major purpose for building basements was the provision of such a storage space. Presently, storage of fuel oil or propane gas is more common, typically occurring outside the building and posing a site design problem.

Use of the sun for heat often means the use of some device for storage to ride out the times with no sunlight. This may pose the same basic space problem as for the storage of other energy sources, unless the building itself is used as a major thermal inertial mass.

3.9 SOLAR SYSTEMS

The relation of buildings to the sun affects design of the construction in various ways. The sun is at once a vital source of energy and a hostile presence for glaring light and solar heat gain. In some ways the trick is to use the energy source without getting the bad effects—not always possible, but a fascinating balancing act for the designer.

3.9.1 Solar Heating Systems

As a heating source, the sun offers the principal advantage of being free of charge. However, obtaining access to sunlight and making constructive use of the source may take some doing. Heat may be obtained by simply soaking in the sun—a basic way of using the thermal inertia principle. Store up heat in some mass (the building?) and use it later when you need it (in the cool of the night).

Heat can also be stored in some medium, such as water, which is circulated in a solar collector and from which the heat is then extracted for use somewhere else. This can also build up a thermal mass in the form of a large tank of heated water.

The sun can also be made to heat up air, with a circulation of the air induced by the natural fluid flow of the heated air. This circulation can be used to move the heat to other locations or to achieve ventilation.

All of these methods have ramifications in the design of the building construction: to accommodate the piping and other elements of systems, to insulate against the loss of heat, to provide a thermal mass (weight) for storage, or to otherwise accommodate the system operations.

3.9.2 Power Generating Systems

Sunlight can be made to generate electricity for use in buildings. The devices and systems for doing this are quite expensive in relation to the power derived, so this is mostly done only in special circumstances: in space, or in some remote places where other energy sources for electrical generation are not available. Usage is most likely for low-powered systems, such as communication and signaling devices.

This may become a critical construction issue in the future if other energy sources continue to become more expensive, depleted, polluting, and so on. Since the surface of the building is generally exposed to the sun, it may have an impact on what we use for surface finish materials.

3.9.3 Daylighting Systems

Almost every building makes some use of daylight, reducing needs for electrically powered illumination to some degree. This may happen without any special efforts by designers, with sunlight and sky light just naturally illuminating the building exterior and leaking in through the windows. However, with some conscious efforts, additional usage may occur—with some deliberate forming and placement of windows, with use of skylights or roof windows, or with light bounced off highly reflecting surfaces.

A major consideration in this regard is for the building windows, and the whole design of windows must usually consider this potential as well as various other ones.

3.9.4 Control of Solar Effects

Controlling sun effects on a building means dealing with both the hostile effects and the potential useful effects. For the building construction this has various possible ramifications, such as the following:

Insulation of the building enclosure (roof and exterior walls) to reduce solar heat gain

Use of light-colored roofing materials to reflect sun heat

Use of heavy construction if a thermal mass is desired, possibly the deliberate use of light construction to *avoid* soaking up heat

Use of special windows, particularly with special glazing, for their enhanced responses to the sun

4

Systems

System is a much overused word, but generally it describes any group of component elements that exist in some ordered relationship. Most often the group is composed to relate to some specific task. A building as a whole entity can be described as a system, although it is more common to use the term to describe some subunit, or subsystem, of a building.

For considerations of the building construction, we visualize the following major subsystems:

The basic building construction, forming the enclosure and divisions of interior spaces

The building structure, performing major load-resisting functions

The foundations, interfacing the building with the site; a subunit of the whole structure

The various service systems for power, lighting, HVAC control, communication, transportation, fire safety, security, and so on

In this chapter the issue of building systems is discussed with reference to the idea of a particular, distinctly definable form of construction. In that context, there are a small number of very ordinary systems that are repetitively used to create buildings.

4.1 HIERARCHIES

Systems with many parts often have the parts arranged in some ranked order or hierarchy. A hierarchy may be defined in terms of the relative importance of elements in reference to some specific task or issue. Individual parts may be viewed as major or minor, primary or secondary, and so on. This may relate to considerations of the principal determinants of the system or to the relative significance of parts to design modifications for a specific purpose.

Modifications in the form of design variations may be more easily done when they deal only with secondary or minor elements of the basic system. Thus with the light wood frame, the studs are primary determinants of the system, while surfacing is secondary to the basic system. Many types of materials may be used for surfacing—effecting major changes in the appearance and various properties of the construction—without essentially altering the basic nature of the structure.

The quality of adaptability just described for the light wood frame is one of the reasons it sustains its popularity; it can indeed adjust to almost endless variety in terms of application to different design styles and situations. There are a very few enduringly popular, basic construction systems that enjoy this quality.

Hierarchies are sometimes exploited for design or construction purposes. One such purpose has to do with the potential for modification and its relation to the relative permanence of elements of the system. For example, with electrical power systems, modification is most easily achieved at the lower end of the system—that is, at the final power delivery point. Modifications at the top of the hierarchy—where the electrical service enters the building and before it begins to branch out—are less easily achieved.

Design of the basic elements of the electrical system usually allows for the ease of modification as just described. However, the installation of the electrical elements within the building construction should also carry out this concept. In a similar way, this idea also applies to the plumbing, HVAC, communication, and other systems. Efforts to achieve this as an integrated design should recognize the hierarchies in all the building subsystems involved.

4.2 DESIGN PROCESS AND TASKS

A major early design task is the selection of the basic systems of construction for the building in general and for its major elements: roof, walls, floors, and so on. Options may be innumerable, or they may be shortened considerably by various design criteria, such as the construction budget, time available for design and/or construction, building code requirements, or strong client preferences. However determined, the firm choice of basic systems at an early design stage will greatly simplify the progress of the design work.

Early decisions must also be made regarding all the various subsystems. This is especially so where major integration of systems is required. Thus the space needs of the HVAC duct system must be anticipated very early, so that its accommodation in the interior space planning and details of the construction can be developed.

With basic systems defined, and their hierarchies understood, the forms of design modification that are most easily achieved can be explored. If the major design goals cannot be achieved within practical limits of the basic systems chosen, it is best to find this out very early. A shift in the basic systems has major effects on the progress of the work, and the earlier it happens, the better.

Although real design work seldom proceeds in so orderly a fashion, the following is a logical process for determination of the construction during the general development of a building design project.

1. Determine the basic systems. This means the selection of general forms and materials (masonry bearing walls with CMUs, for example).
2. Determine specific features of the basic elements (mortar type, basic unit type and dimensions, typical details, and other critical specifications for the CMU walls).
3. Consider effects on major building elements. This anticipates various problems of design integration. (How do unit dimensions affect window and door sizes, room dimensions, etc.?)
4. Develop all necessary major architectural details for use of the system. This tests the variability of the system and the ability to generate the range of situations that must be accommodated in the building, as well as the potential for alternatives.

Above all, the design process must continuously be an informed one. Designers of all the individual subsystems must be kept abreast of the development of other systems that affect their work in a major way. Each designer should be aware of the major implications of his or her own work as it affects others.

4.3 MIXTURES

Buildings consist of large collections of individual parts and subsystems. There are very few manufacturers of whole buildings; in the main, each manufacturer produces only selected parts. It is the task of the building designer to determine the collection of parts for a particular building.

Practicality generally dictates that the parts of buildings are mostly obtained as manufactured products, selected from the catalogs or samples of the products as supplied by manufacturers or suppliers. In this situation, the building designer has the major task of comparative shopping and mixture making.

While a building represents a giant mixture, it is usually a mixture of subsystems and components, each of which may also be a mixture. For example, a designer may choose to use a light wood frame (2 × 4 wood studs, etc.) for a wall. This is a well-defined basic construction component and it can be "mixed" with various other components for an appropriate building assemblage. But the stud wall itself is a mixture, with studs produced by one maker, surfacing by another, connectors by a third, and so on. Each item can represent a design choice not necessarily implied by the general choice to use a wood stud wall system.

Selection between the available products for a single use (surfacing for a stud wall, for example) requires comparative value analyses, and individual suppliers cannot be fully relied upon to provide unbiased assistance. The larger view of options, as well as real freedom of preference, is required. One type of wall surfacing may be the best in its class, with many superior qualities, but those qualities may be of minor concern for a particular application, and a lower-quality, less expensive product may be perfectly adequate for the limited task at hand.

Designers (shoppers, buyers) must look diligently for the best choice for each individual part for a building. But the whole building is more important than any individual part. The best roofing material, placed over a poor roof structure, will not produce a good roof.

4.4 APPROPRIATENESS OF SYSTEMS

The achievement of good construction begins with the choice of the appropriate materials, products, and systems for a particular use. Add in careful detailing, proper and thorough specifications, and competent construction work, and a well-built building just might result. But a well-built building in the wrong place or for the wrong use will still be inappropriate.

It is one thing to put plaster on a wall in the right way; it is another to select plaster as the appropriate surfacing for the wall for the situation at hand. Product information and standard specifications can deal with the correctness of the plaster itself, but the larger question of whether or not to use plaster is in the hands of the designer.

Manufacturers or industry associations sometimes provide guidance to the appropriate use of their products. However, they have in general a major interest in promoting the use of their products, so they do not much like pointing out all the situations in which those products are inappropriate.

Building codes provide some guidance for the selection of materials and systems for appropriate situations. This occurs mostly as a negative form of criteria—prohibiting certain uses in particular situations. These criteria should be determined for any design work and used to narrow the choices very early on, but they typically leave room for many options.

Various unbiased reference sources exist for some assistance in making appropriate choices. However, individual buildings always have some specific qualified conditions that make simple evaluations difficult. Timeliness is also a major concern, as well as the many influences of regional situations.

In the end, the observations, experience, and judgment of the designer must prevail. This is what design is all about, and the task is humbling.

4.5 CHOICES: THE SELECTION PROCESS

Design of building construction is usually simpler and faster to achieve when choices can be made in large bites. Choosing preestablished (or essentially predesigned) systems is one means of doing this. If the choice is good in all regards, the size of the bite is irrelevant.

Preestablished systems offer the potential advantage of some demonstrated success. If the demonstrated successful cases have a reasonable match with the situation for a proposed design, all the parties concerned—designer, builder, owner, investors, insurers, and code-enforcing agencies—may take some comfort in an assurance of success. This is, of course, the actual case, making any really innovative design difficult to sell in many quarters. Designers should not be totally discouraged, but must understand the realities in achieving any significant changes in the "good old ways" of doing things.

Exciting, uplifting architecture can really be achieved with very ordinary means of construction, but can also—on occasion—come from real innovation or dramatic usage of the construction. In either case, the designer usually has a considerable understanding of, and appreciation for, the construction details and processes. If ordinary means are used, the designer understands their limits and potential, as well as what significant architectural design variations can be achieved within the basic systems. If ordinary means are rejected, the designer usually does so in full knowledge of what exactly is being discarded and what the needs are for any replacements. Otherwise, innovations will inevitably be naive and quite likely unsatisfactory in some performance aspects.

4.6 SYSTEM DEVELOPMENT CONCERNS

General planning of construction as well as choices for basic systems for an individual project must be made with the whole development of the project construction in mind. Certain portions of the work may be accomplished in a relatively separated activity for various reasons. One of these reasons is the separate functions represented by different design professionals.

Traditionally, architects were generally the master planners and coordinated the design work of most building design projects. This is still frequently the case, but the emergence of the separate activity of construction management and the role of developers has in many cases moved the command decision level to other people. In addition, the professional stature of many other members of the whole building design team has improved and strengthened. Interior designers, landscape architects, specification writers, environmental planners, and others have advanced the status of their professions, in many cases at the expense of some erosion of the leadership stature of architects.

In any event, the work of many strongly independent design groups must be coordinated for a successful project. In addition, the fractional division of the construction work is also a factor. This division often results in the necessity to literally produce separate design documentation. Any project of significant size requires architectural drawings, structural drawings, electrical drawings, and so on, thus creating a simple and necessary—but not easy—job of coordination of the multiple views of the design.

Continuing the problem is the fractional division of the building industry, with most manufacturers devoted to single materials or product lines. Thus design data, product detailed information, costs, availability, certifications, and other matters must be separately pursued. Industry organizations and standards are mostly quite narrowly defined and even regional in many cases.

The degree of fractionalization eventually overwhelms many attempts to coordinate the work—both for the design and the construction. This is a major management problem and designers should be aware of it, but they should also realize that there are few *design* solutions for it. With regard to design, however, there are a few key relationships that should be given attention to be assured that an integrated design effort is achieved. These are treated in the following discussions.

4.7 FUNDAMENTAL NEEDS OF SERVICES

It helps if the general designers of the building construction understand some of the primary needs and limitations of the various building services. For example:

> Waste drainage piping (from sinks, toilets, etc.) must facilitate gravity flow of liquids. This is especially a problem for long horizontal runs. A single pipe may theoretically fit inside some interstitial space, but its position must change if it has a long horizontal run.
>
> Waste drains must be vented—usually through the roof and as directly vertically as much as possible. This is a problem for the roof, as well as all the construction between the vented fixture and the roof.
>
> Hot-water pipes and heating ducts get hot; chilled-water pipes and air-conditioning (cooled-air) ducts get cold. Surrounding construction can be heated or cooled. Thermal expansion and contractions can create problems. Chilled objects can sweat and create moisture problems.
>
> Electrical wiring must frequently be altered, and various other elements should be accessible for adjustments, repairs, alterations, and so on. Embedding elements in the permanent construction can create many problems.
>
> Equipment with moving parts—notably motors, fans, pumps, and chillers—create both noise and direct, physical vibrations that can be annoying. They should be isolated—possibly by good planning of their locations, but also by use of vibration isolators, sound-insulated partitions, or other special construction.

As noted previously, compromise is the name of the game. Ideally, all building systems should operate or exist with optimal conditions, but economic feasibility, technical limitations, and some priority of design values and goals must be recognized.

4.8 OPTIMIZATION AND SUBOPTIMIZATION

An idealistic design goal is often to optimize each subsystem and each individual element: to strive for having the very best doorknob for its purpose, for example. It is truly useful to visualize this and to establish the real criteria for making such value judgments, but true optimization is quite elusive.

A frequently occurring scenario consists of having some selected subsystem perfected within its own limited design reference, but causing real problems for other units of the overall system within which it functions. Thus a really efficient structural system (lightest, cheapest, most fire-resistive, etc.) may shine in its own right, but may not accommodate the placement of ducts, piping, and wiring, the embedment of built-in fixtures, or the attachment of finish materials. This is a classic example of *suboptimization:* the shortsighted perfection of a single unit without realization of its impact on related units.

The management of design for complex systems (with systems of systems) requires the constant recognition of an ordered set of values, in which the interests of the whole system are predominant. The cost, time for construction, energy use, and so on, for the *whole* building is generally more important than it is for any subunit.

4.9 MULTIPLE-USE ELEMENTS AND SYSTEMS

Many parts of the building construction have singular uses, for example, light switches. On the other hand, many parts also have multiple uses. Windows, for instance, are used for viewing out, for viewing in, for obtaining natural daylight for illumination of interiors, and sometimes for ventilation.

Design of each individual element must recognize all the potential functions or usages for that element. Thus the designer of the building exterior—for whom the windows are principal design elements—must recognize that the windows may be key elements for building security, for building lighting, or for the general HVAC control.

Multiple usage may sometimes be implemented for design efficiency. For example, interior walls that are necessary for space division may also be used for vertical structural support, for fire separation, and for the housing of wiring, piping, and ducts. This may represent a design efficiency by reducing need for separate elements to achieve these other purposes: structural columns, separate fire barriers, duct shafts, and so on.

However, multiple usage sometimes impedes modifications that may be important to facilitate. In many occupancies, such as office buildings and retail stores, it is often highly desirable to be able to make frequent rearrangements of interior walls. In this case, the less a wall does—besides just be a wall for space division—the better.

4.10 PASSIVE USAGE OF CONSTRUCTION

For the achieving of various aspects of control of interior environments, an important element in the control system often is the building's construction. This form of contribution to the control is generally characterized as being a *passive* one, by comparing it to *active* contributions made by installed, mechanically operated systems.

For purposes of contributing to interior conditions relating to control of lighting, ventilation, temperature, sound, fire safety, or security, any proposed form of construction can be evaluated—often with specific criteria, producing a rating that can be used for comparative analyses. These ratings are routinely utilized in value judgments made in building design procedures.

Contributions of the basic contruction are typically included in the design process for the active—mechanical—systems. Thus the choice for a furnace is based on an estimate of the total heating load required, which is in turn based on an analysis that includes the thermal insulative effects of the roof and exterior walls.

This is a major area for integrated design efforts. Obtaining a particular interior condition (for example, 68 degrees Fahrenheit on a cold day) may be achieved by some combination of effective insulation of the enclosure and operation of the heat-generating system. Comparisons can be made of alternatives that manipulate one or the other: the passive or the active systems. Add more insulation or increase the size of the ducts and furnace; which is cheaper?

It is important for building designers to understand the general aspects of the passive contributions made by the basic building construction. This establishes a level of responsibility in making decisions about choices of materials and details of the construction, beyond the simple concerns for appearance and proper construction. No person should make judgments for these choices who does not fully understand the implications of the various ratings for the construction in terms of hours of fire resistance, transmission factors for thermal flow or airborne sound, and so on.

4.11 CHANGES IN NEEDS WITH TIME

A peculiar requirement of modern buildings is their general need to facilitate modification. The dynamic nature of our society makes this imperative. People relocate regularly; businesses reorganize; the urban fabric continuously changes—growing, decaying, regrouping. Rapid changes in technology bring many sudden demands for facilitation of new opportunities in terms of availability of new materials, furnishings, and equipment for buildings.

Playing against this rapidly shifting scene is the great expense of building costs, which makes the writing off of initial construction over a long period mandatory for most investors. The net effect is that buildings must be built for permanence but somehow be relatively easily modifiable.

Ease of modification is a design goal and involves both basic planning and the final decisions for choices of materials and details of the construction. It also needs to permeate the whole design effort, including that of the structure and all the building services where possible. This is relatively easy for some things, such as wiring, ceilings, and plumbing fixtures; it is not so easy for sewers, foundations, shear walls, or major elements of equipment for basic services.

Designing for special requirements, such as ease of modification, may require additional effort and extra construction costs. In some ways, however, it merely means picking between competing alternatives with a view to those that facilitate ease of change. Bolted joints can be undone, for example, while glued or welded ones cannot. Surface-mounted wiring can be relatively easily modified, while wiring buried in the permanent construction may not be possible to rearrange.

Modification is especially important to certain occupancies, such as office buildings or retail stores, and is typically given high priority in design for such uses.

4.12 ADAPTATION

Forms of construction for both the basic building shell and its various service systems can be visualized in terms of relative adaptability to change. For example, wood and steel structures are generally more adaptable to modification than those of cast concrete or masonry. Ceilings consisting of modular systems of frameworks and loose units are quite easily rearranged, while plastered ones must generally be fully reconstructed.

In making choices for materials and forms of construction, it is important to consider the relative importance of adaptability to modifications, involving minor rearrangements, complete relocation, extension for growth, and so on.

Adaptability has many aspects, ranging from the fully demountable and completely rearrangeable structure to surfaces that can facilitate various finishes. The former is quite difficult to achieve, except for a few special structures; the latter is a common property for many interior surfacing systems. It is important for designers to consider the nature of use of buildings in terms of ongoing demands for change, refurbishing, maintenance, and so on. This is a major part of consideration of building design as being one for a long time of use, as opposed to design of stage sets or sales window displays. The initial, cleanly executed construction may look great for the photographs for the designer's portfolio, but it is only Day One for the building users.

4.13 RETROFIT

The best designs, based on the best available information, cannot fully anticipate the future. New times bring new concerns, new technical capabilities, new uses for buildings, and newer and more data-based design processes. The inevitable result is that any new building design begins to age as soon as it leaves the design office. Updating of buildings (retrofit, renovation, whatever) goes on even in the process of initial construction.

This is a major concern for the design effort to facilitate modifications, as it often goes to major components of the basic systems. If a whole wall is wrong, it is more serious than just refinishing the interior surface. If the basic form of the heating system (steam radiators?) is wrong, it is more serious than just using a better furnace or a different fuel for heating.

Retrofit is currently a big part of the building design and construction fields, as pressures increase for continued use of older buildings. The lessons learned from this activity will hopefully influence the design of tomorrow's buildings, making them generally more adaptive to future retrofit.

4.14 INTEGRATION OF BUILDING SERVICES

Most buildings have electrical power, lighting, water supplies, waste piping, and HVAC systems. Multistory buildings have elevators, escalators, and vertical shafts for wiring, piping, and ducts, chimneys, incinerators, mail chutes, and so on. Operation of the various services requires equipment that must be housed inside, on top of, or under the building.

Incorporation of service elements is a major chore in interior design, requiring considerable cooperation between the designers of the various separate systems. Overall coordination generally falls to the architectural designer and must be accomplished along with the design of the general interior construction.

A major consideration that must be made in this effort is the avoidance of structural interference. This involves the problems of inserting all the nonstructural elements (including nonstructural construction) in a way that does not critically disturb the required functioning of the building structural system. Of typical concern are holes cut through walls or webs of beams, floor and roof openings that disturb orderly layouts of framing, and need for horizontally continuous spaces that preclude locations for columns, bearing walls, shear walls, X-bracing, and so on.

The structure must function for safety of the building. If it cannot do so adequately and accommodate the other necessary building functions, something has to go. The structure is not inviolate, and some forms of compromise are to be expected. This is a two-way street, and a lot of judgment is required.

5

Case Studies

The purpose of this chapter is to present a number of examples of building types, construction usages, and site situations and to illustrate the development of the general construction for each example.

The presentations here are meant to complement the materials in preceding chapters. Critical isolated issues are dealt with more extensively in those chapters, and that material should be used as a resource for the more detailed information on individual items shown in these examples.

However, designers must ordinarily face the problem of the whole construction, even though the actual design may be worked out by concentrating on one problem at a time. Here, the presentations begin with consideration of the building's general form and construction system, followed by presentations of some of the primary increments of the construction with all of the significant parts displayed. The discussions then proceed to considerations of major service systems, with emphasis on factors that significantly affect the development of the building's construction.

The building designs presented in this chapter are not meant as illustrations of superior architecture. The interest here is directed mostly to setting up reasonably realistic situations that cover a range of circumstances in construction design. It is not the intention to present boring, ugly buildings, but the use of a limited number of examples to display a broad range of construction use is the dominant concern.

An attempt has been made to show construction usage and the form of details that are reasonably correct. This is, however, a matter for much individual judgment and is often tempered by time and location. There are not many situations in which only one way is correct and all others are unequivocally wrong.

Reasonably correct alternatives are often possible, and the particular circumstances for an individual building can make a specific choice better on different occasions. Differences in climate, building codes, local markets, building experience, or community values can produce a way of doing things that is preferred. Over time, changes in any of those areas of influence will produce variety and, it is hoped, some evolution toward a better way of doing things.

The concentration in this book is on building service systems and their relations to the general building construction. General form and construction details of the buildings are presented here only to the extent necessary to understand the basic development of the construction, with emphasis on those parts that especially relate to design and installation of the service systems.

The more complete development of the construction for these examples is considered in the two books which are the first volumes of this series—*Building Construction: Enclosure Systems* and *Building Construction: Interior Systems.*

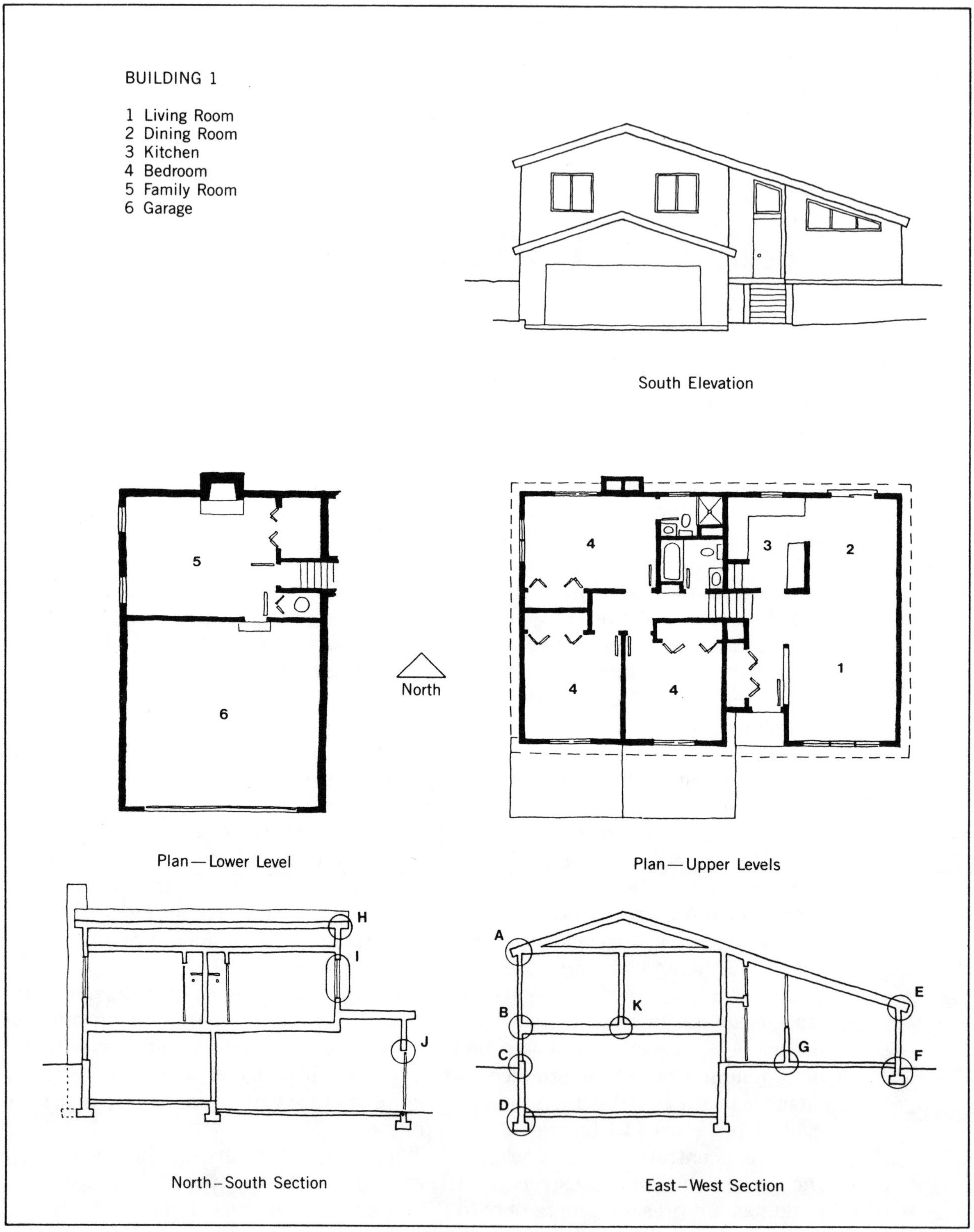

Figure 5.1 Building 1.

Some of the details of the buildings that are developed in those books are shown here for reference, but are not fully discussed in their own rights in this book.

Finally, although incorporation of building services is a major concern for the development of the building/site interface, the full development of site design considerations is not treated here. These issues are treated in depth in the third volume of this series, *Building Construction: Site and Below-Grade Systems.*

5.1 BUILDING 1

Single-Family Residence Light Wood Frame Sloping Site

This is a building of modest proportions, not impressive to architectural critics, but a castle by world housing standards. As a type of building—and in this general size and general character—it is probably the most widely constructed building in the United States. Cape Cod cottages used this basic structure hundreds of years ago and are still imitated as a style in extensive housing developments. The form of house shown here, in Figure 5.1, is that of a typical suburban split-level, but many details are really reminiscent of the basic Cape Cod cottage.

Figures 5.2 and 5.3 show the details for the general building construction. As shown here, the light wood frame construction system provides considerable hollow, interstitial space that can be extensively used for installations of out-of-view piping and wiring and some built-in elements of the various service systems. Some elements of the latter type that are routinely accommodated in this fashion are power outlets, light switches, wall and ceiling HVAC registers, and some recessed ceiling light fixtures.

There are two situations shown in the construction details, however, where burying of service elements is somewhat of a problem. Details C, D, and F show the concrete slab floor and foundations. These solid concrete elements offer less accommodation to burying of elements, although some wiring is occasionally so treated, as well as small items such as power outlets.

Another situation of this type is shown in detail E2 in Figure 5.3. With the plank roof deck, any elements for services at the ceiling must likely be surface-mounted or suspended, whereas the use of the light wood construction shown in detail E1 would permit relatively easy concealing of any wiring, recessed lights, or even small ducts.

Figure 5.4 shows the general scheme for development of an all-air HVAC system for building 1. There are several possible variations on this system, with certain forms more appropriate for particular climatic conditions. Shown here is a system that favors air conditioning (cooling) through use of high wall registers, a system more appropriate for warm climates where heating is not a major concern.

In cooler climates, especially if air conditioning is not developed by the central air-handling system, it would probably be better to use a system with perimeter registers in the floor. For the construction, this would be relatively easily achieved within the light wood frame for the upper-level bedroom portion of the house. However, the floor registers and feeder ducts for the portions with slab floors would be somewhat more difficult. This might favor the use of crawl space construction for the living room portion of the house. Use of the crawl space would also make it somewhat easier to assure warm floors and a better response to the problems of edge heat loss at the ground level.

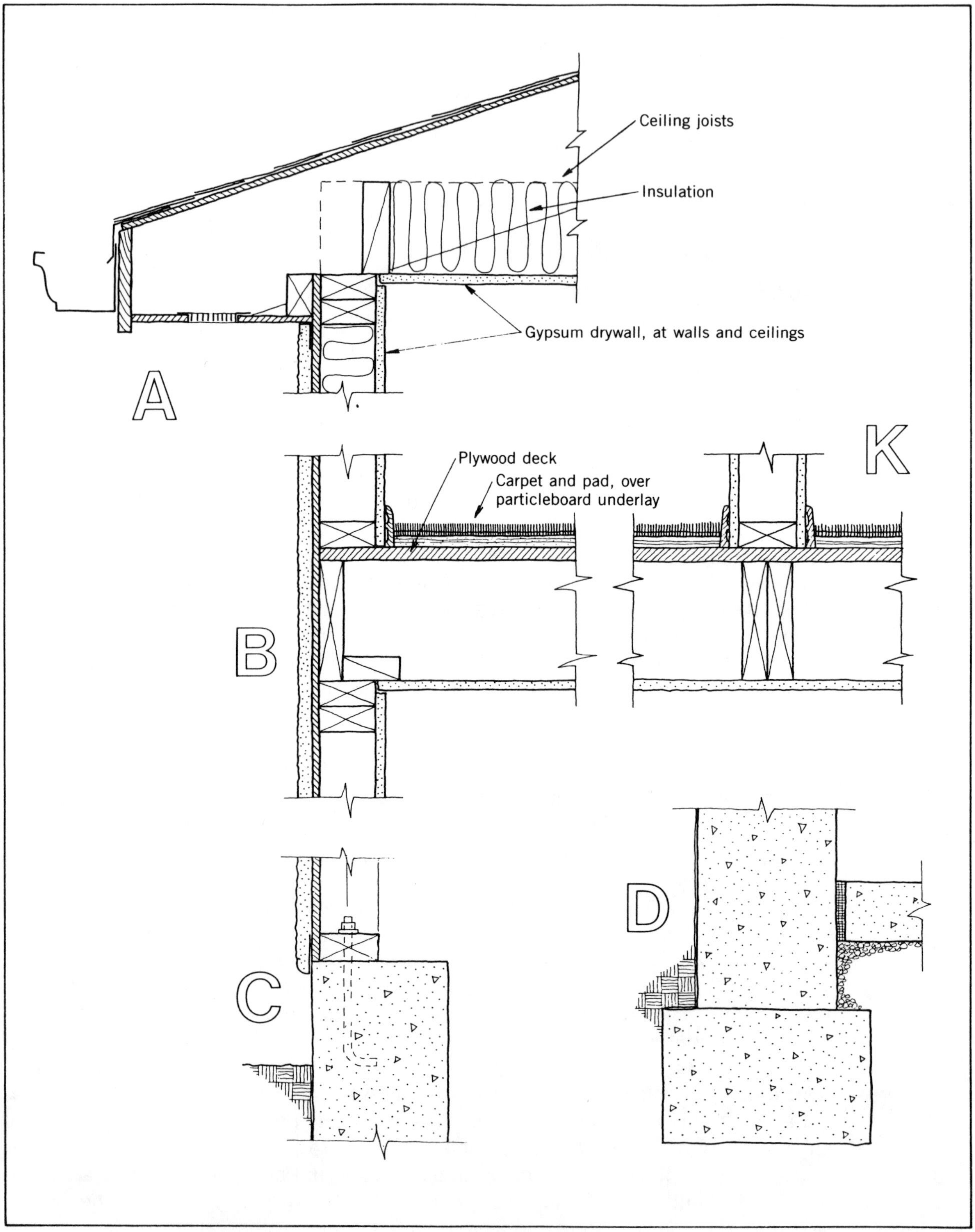

Figure 5.2 Building 1: general details of the building shell.

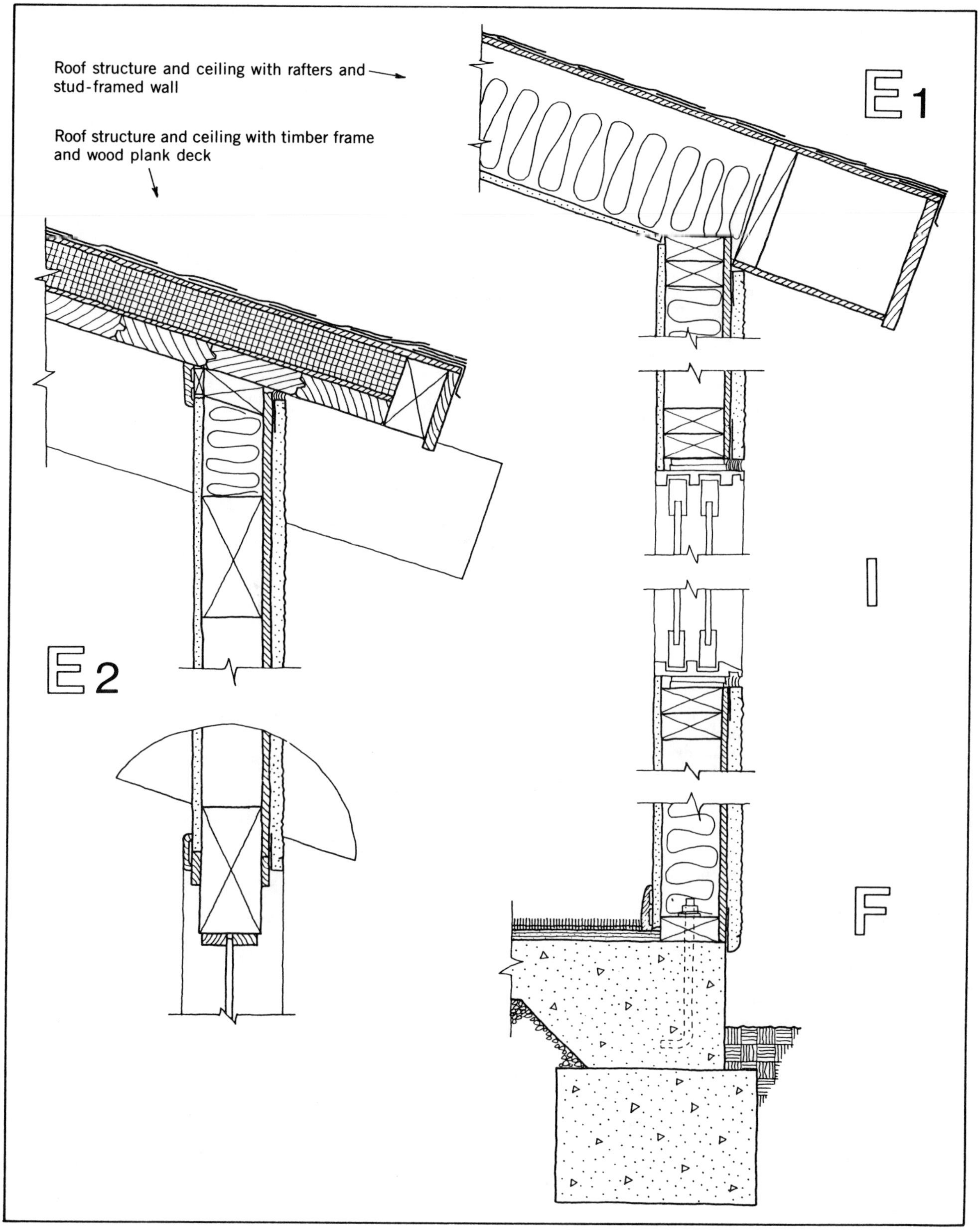

Figure 5.3 Building 1: general details of the building shell.

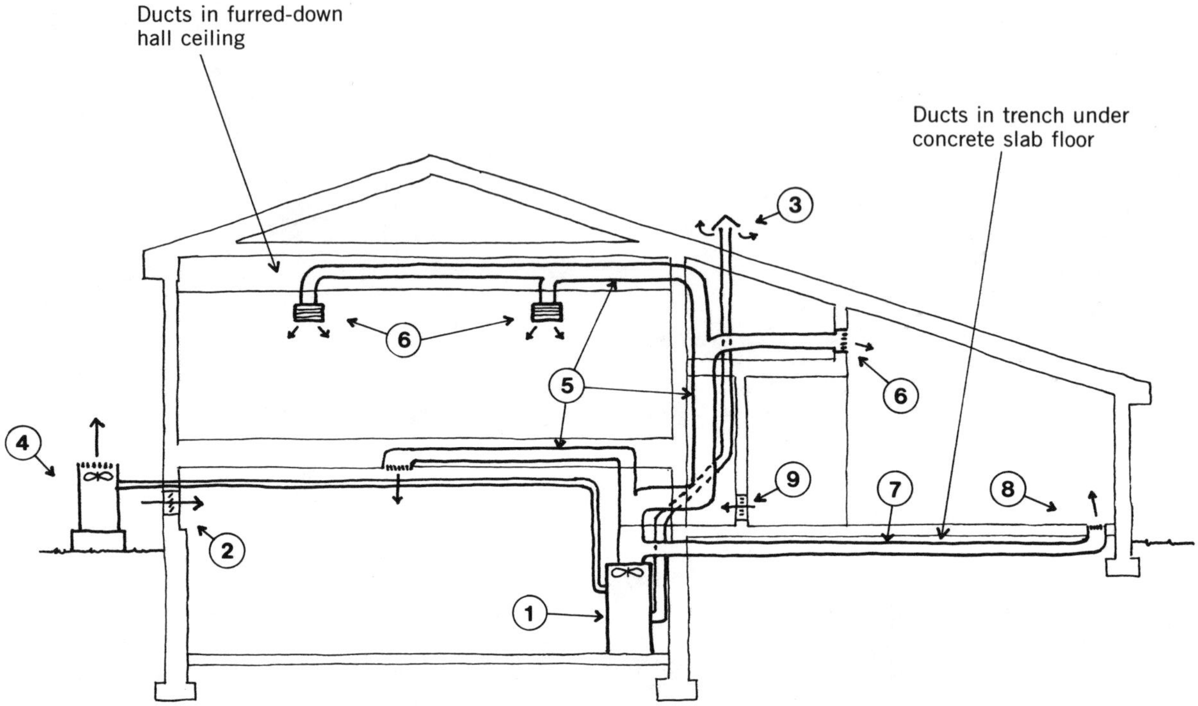

1 Central HVAC Processing Unit
Fan, furnace, air-conditioning chiller, filters, etc.; needs hookups to electric power, supply and exhaust ducts, signal wiring from thermostat, chimney stack, coolant line to outdoor unit

2 Fresh Air Intake
For combustion in furnace; to general basement space or direct to furnace

3 Chimney
Exhaust of fumes from furnace

4 Heat-Dissipating Unit
Component of central air-conditioning system; needs outdoor location

5 Conditioned-Air Supply
Ducted supply of conditioned air to individual spaces in house; discharge through registers in ceilings, walls, or floors

6 Wall Registers
Accessible for adjustment; high wall placement to throw air across rooms; work best for air conditioning (cooling)

7 Underfloor Ducts
Ducts in trenches under floor slab; feed to floor registers, usually at building perimeter

8 Floor Registers
Work best for heating; warm cold outside walls and windows; could also be baseboard units

9 Air Return to Central Unit
Probably one centrally located wall grille; preferably short distance from central unit

Figure 5.4 Building 1: general scheme and details for the HVAC system.

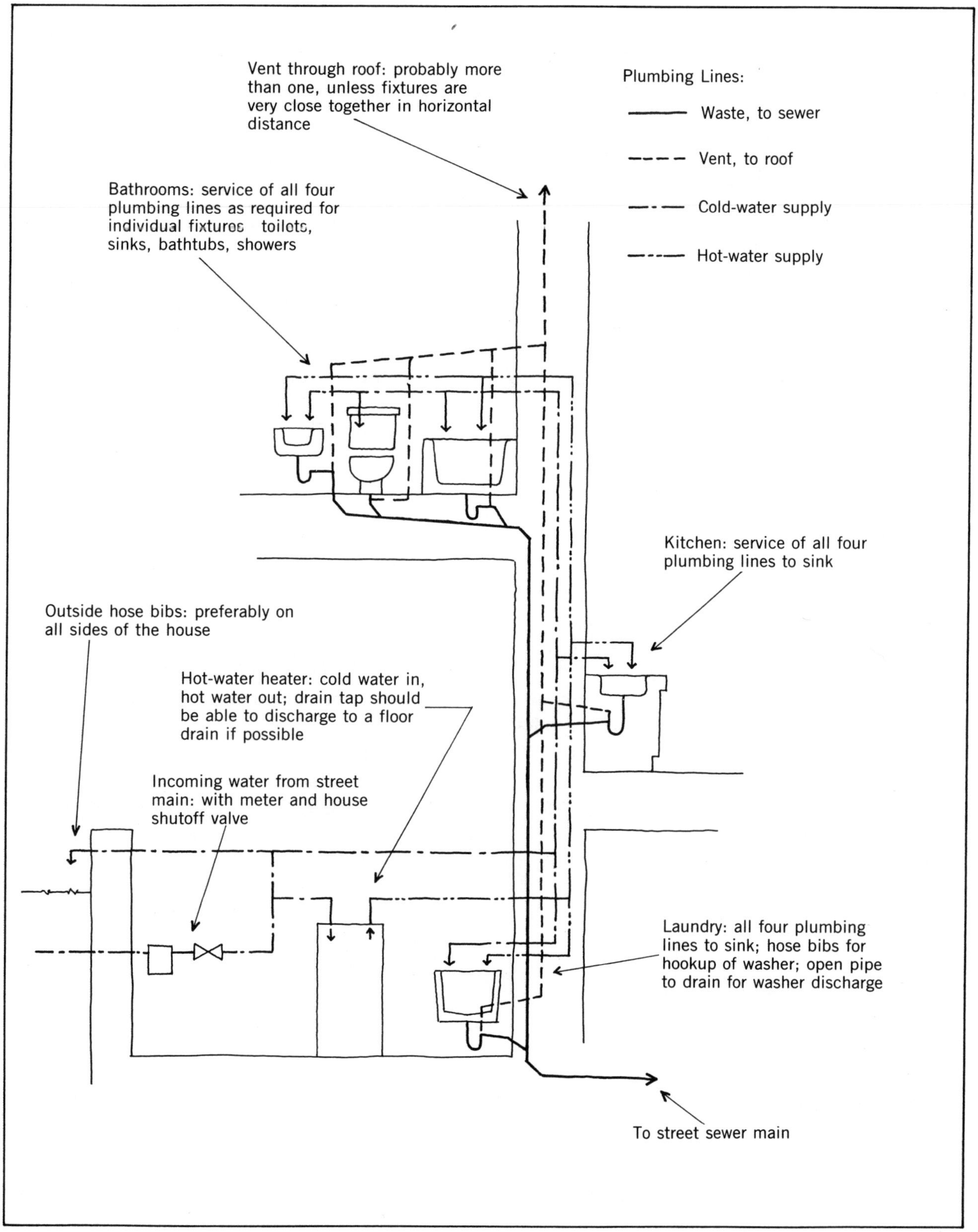

Figure 5.5 Building 1: plumbing systems and details.

Separately developed—although with some similar construction details—are the ventilation exhaust systems for the bathrooms and kitchen. Use of the dropped ceilings in these areas greatly simplifies these installations.

Figure 5.5 shows the general nature of the plumbing systems for water supply and waste. Again, the light wood frame is reasonably accommodating for these installations. In fact, it is almost too easy to hack up the light wood frame, resulting in some serious losses of the structural frame in some cases. Generally, the structure has considerable redundance and load sharing by the closely spaced members, but some careful inspection should be made to assure the structural integrity of the frame after the plumbers and electricians get through with it.

The construction work must be carefully coordinated where the plumbing systems pass through or under the concrete slabs, walls, and foundations. If other services are also delivered underground to the site, these must also be developed in cooperation with the scheduling of other site and foundation construction work. Tearing up cast slabs, cutting holes through concrete walls, or decimating the finished landscape work should be avoided. The construction work of several different trades and subcontractors and the design work of several different professionals may be involved. This is where the architects, construction managers, or general contractors earn their money, if serious conflicts are avoided.

A likely additional plumbing element is a perimeter footing drain for the lower portion of the house where the floor is below grade. Roof drainage downspouts (vertical leaders) may also be fed into this system, if good site drainage away from the below-grade walls cannot be assured.

Some details for the basic electrical power system are shown in Figure 5.6. The typical system for a residence of this size would provide for both 110- and 220-volt service. The 110-volt system is used for lighting and the general outlets providing for random hookup of movable items, such as TVs and floor lamps. The higher-voltage system would be used for the HVAC system, clothes dryers, electric range/ovens, and possibly for the attachment of heavy shop equipment if a workshop is provided.

Small transformers are usually used to step voltage down for built-in systems for doorbells, smoke alarms, thermostats (for the HVAC system), intercoms, and possibly some outdoor lighting. Figure 5.7 shows some details for lighting elements that are built into the construction. Wiring for lighting is integrated with that for the general power distribution, while the wiring for low-voltage systems (including phones and cable TV) is separately, and somewhat more casually, developed in the construction.

Figure 5.8 shows some details for the various low-voltage electrical systems. Altogether, there is a lot of wiring in the average, simple residence, and the popularity of the light wood frame stems significantly from its abundant interstitial spaces which comfortably accommodate this wiring, as well as the many built-in electrical components and fixtures the wiring services.

5.2 BUILDING 2

One-Story Commercial
Light Wood Frame
Flat Site

This is a small, one-story building with a flat roof and a parapet wall on all sides (see Figure 5.9). Although the building plan as shown is generally symmetrical about a central corridor, the building exterior is developed to relate to a site orientation that

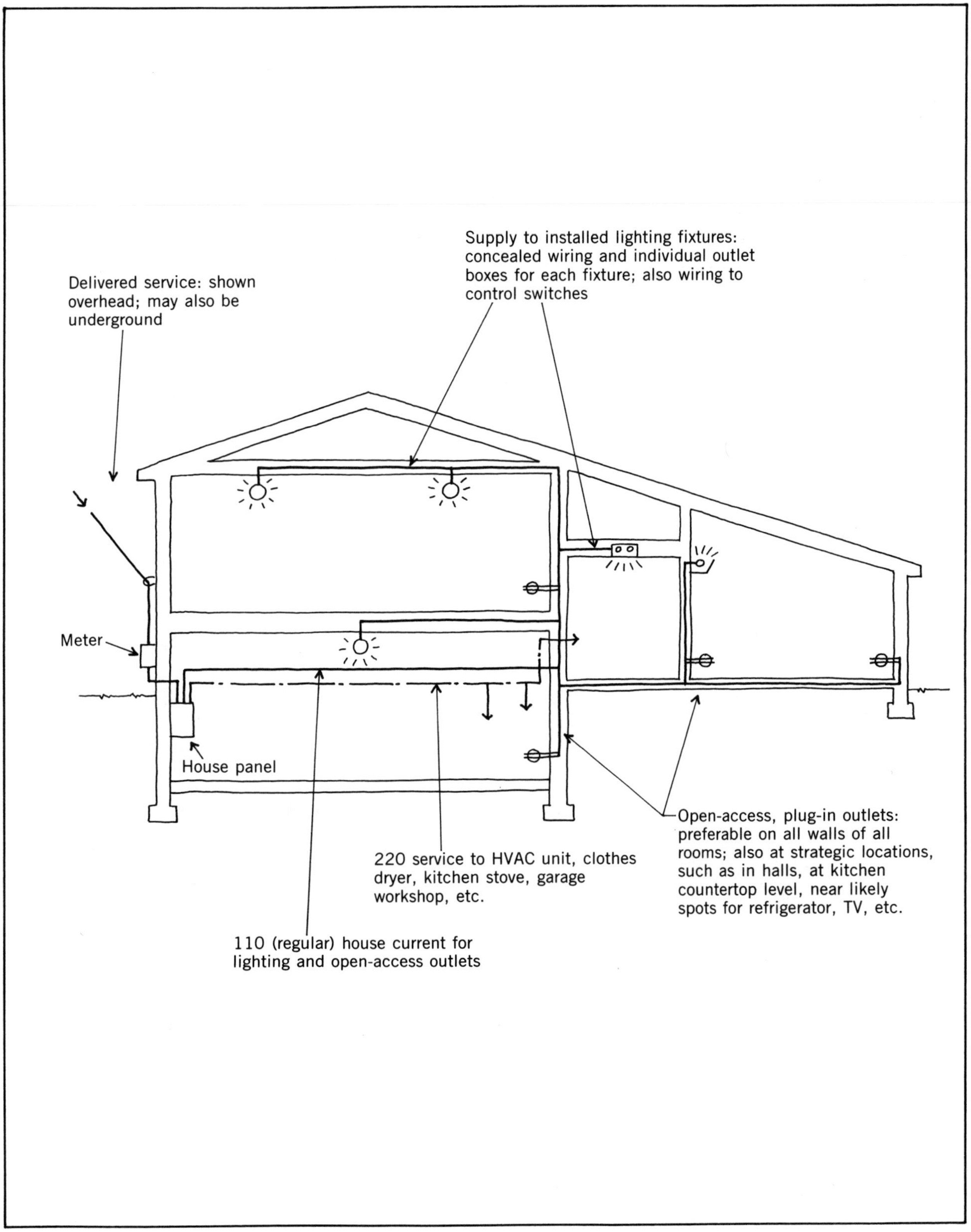

Figure 5.6 Building 1: details of the electrical power system.

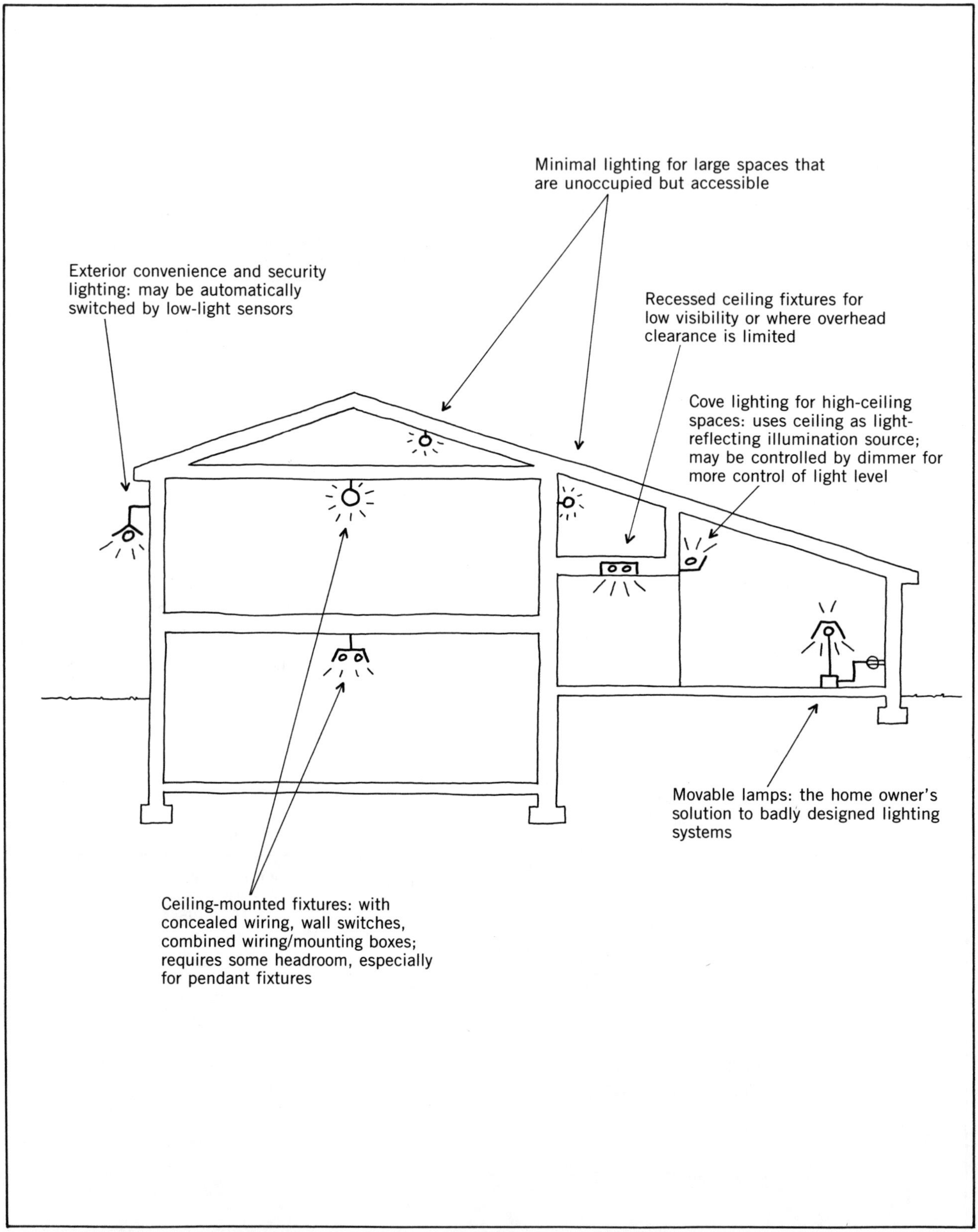

Figure 5.7 Building 1: details of the electrical lighting systems.

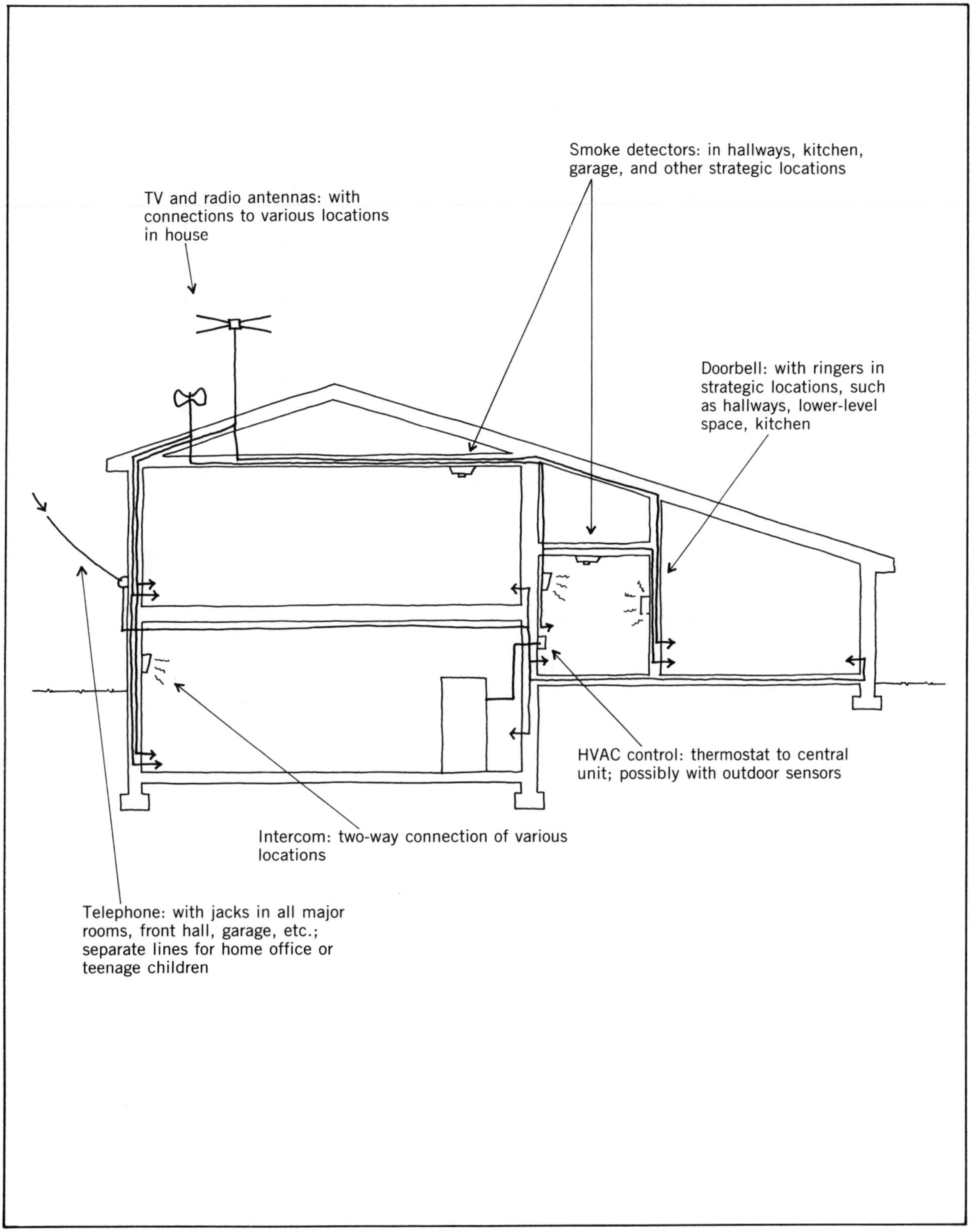

Figure 5.8 Building 1: details of the signal wired systems.

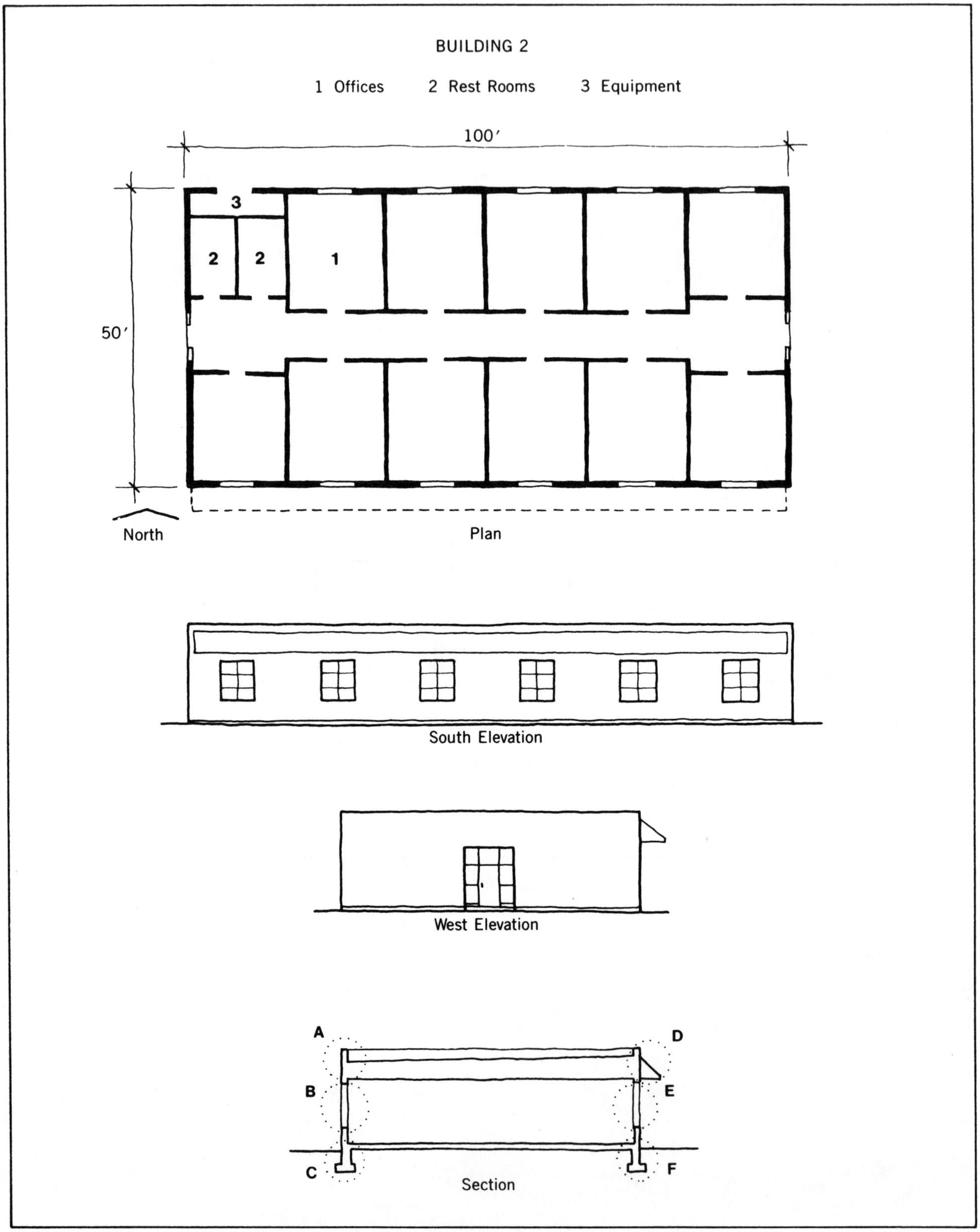

Figure 5.9 Building 2.

results in one long side being treated as the front (street-facing) and the opposite side as the rear. A short cantilevered canopy is provided on the front side and the roof surface is sloped to the rear. The most likely method for draining the roof surface in this situation would be by scuppers through the rear parapet and vertical leaders on the rear wall.

The general details of the construction are shown in Figure 5.10. These reveal the structure to be a light wood frame with structural plywood siding on the rear and ends and brick veneer on the front. The floor is a concrete pavement and the roof structure consists of a plywood deck on closely spaced clear-span trusses.

Use of the clear-span roof structure permits interior walls to be placed at will, with remodeling for other desired arrangements in the future easily achieved. This form of structure is common for commercial buildings built primarily as speculative rental properties, as it allows for a wide range of prospective tenants.

Despite the accommodation of interior rearrangements, certain features become relatively fixed and not subject to easy change. Primary elements in this regard are the building foundations and major structure. Also difficult to change are windows and exterior doors, making some reasonably accommodating locations the most desirable. Plumbing, particularly sewer lines beneath the concrete floor slab, is also hard to change.

Changes will be made more easily if most services are installed in walls or in the space above the ceiling. This will allow for major reworking of the distribution of wiring, ducts, and various items for the building power, lighting, communication, and HVAC systems. The development of details for the construction of both permanent elements and speculatively demountable elements should be made with these provisions in mind.

The prefabricated trusses can be formed to have a flat bottom chord but a sloping top one to accommodate the roof drainage. For the simplest construction, the ceiling may be applied directly to the bottom chords of the trusses. With the open webs of the trusses, this still allows for major elements, such as ducts, to run perpendicular to the trusses within the ceiling interstitial space. This is a case where a deeper-than-average truss might be used to place the ceiling at a desired level or to provide a maximum of space between truss web diagonals.

The basic construction shown for this building is quite typical, allowing for regional differences. This is a small building in a total floor area, permitting the use of the light wood frame by most building codes. However, the construction may have to develop a one-hour fire rating, depending on the occupancy or on zoning requirements. If the plan were larger, the use of a sprinkler system or the placing of a separating fire wall might still allow the light wood frame to be used.

Depending on the type of occupancy, this building might be developed with no ceiling, exposing the overhead roof structure. This is less likely to occur if any interior partitioning is used, as extending partitions into the truss construction is quite messy. The various problems of dealing with exposed, overhead construction are discussed more fully in other building examples in this chapter.

The general form and some details for the HVAC system are shown in Figure 5.11. The system shown is an all-air, single-duct system, using the ceiling interstitial space as a plenum return. The air-handling equipment is installed on the roof. Heating is generated by a hot-water boiler that supplies piped hot water to a fan coil system. Chilled water is similarly supplied by a refrigeration system.

In this building, the water heater and chilled-water systems are housed in a room at floor level, which is accessed from the exterior (see Figure 5.9). The chilled-water

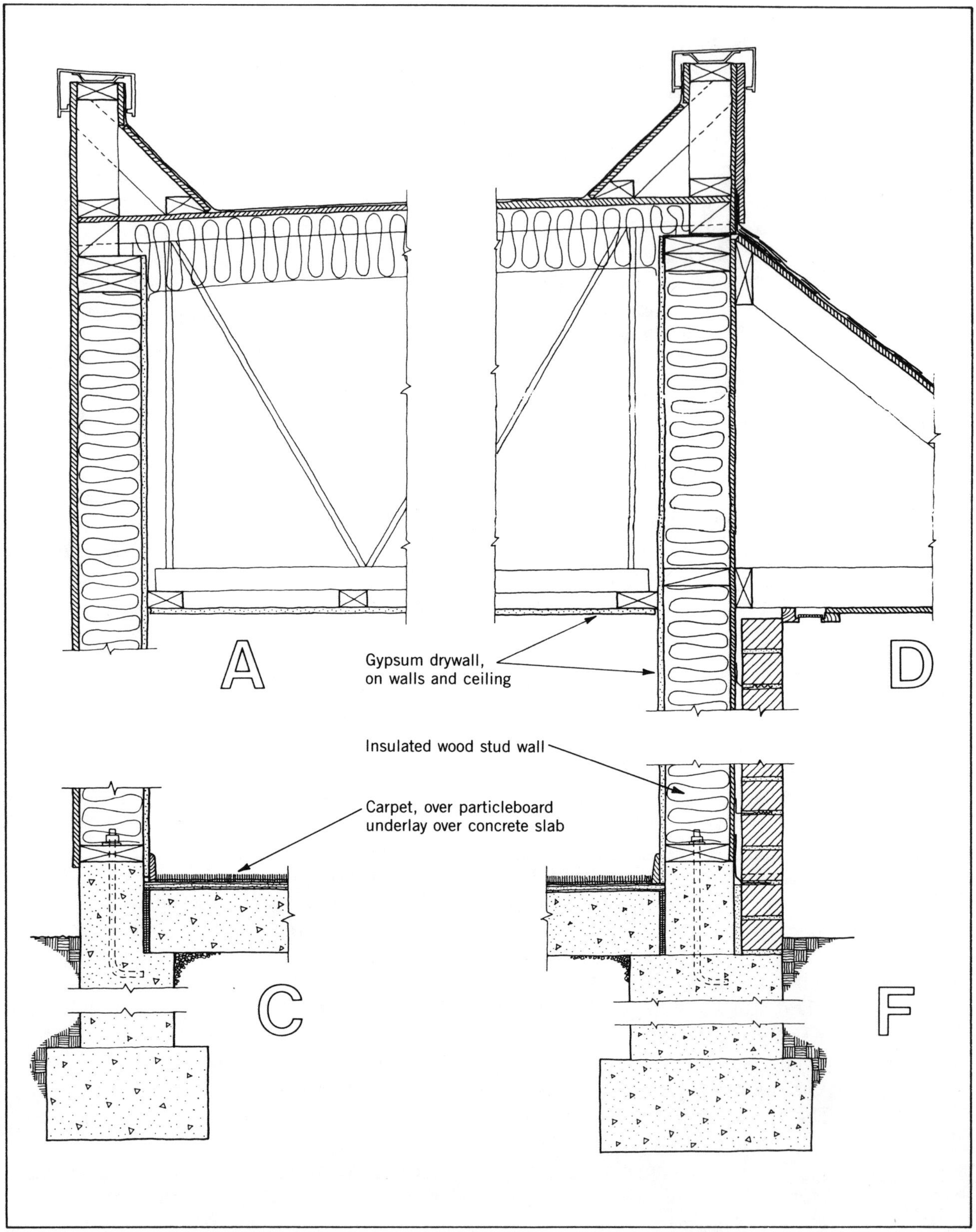

Figure 5.10 Building 2: details of the building shell.

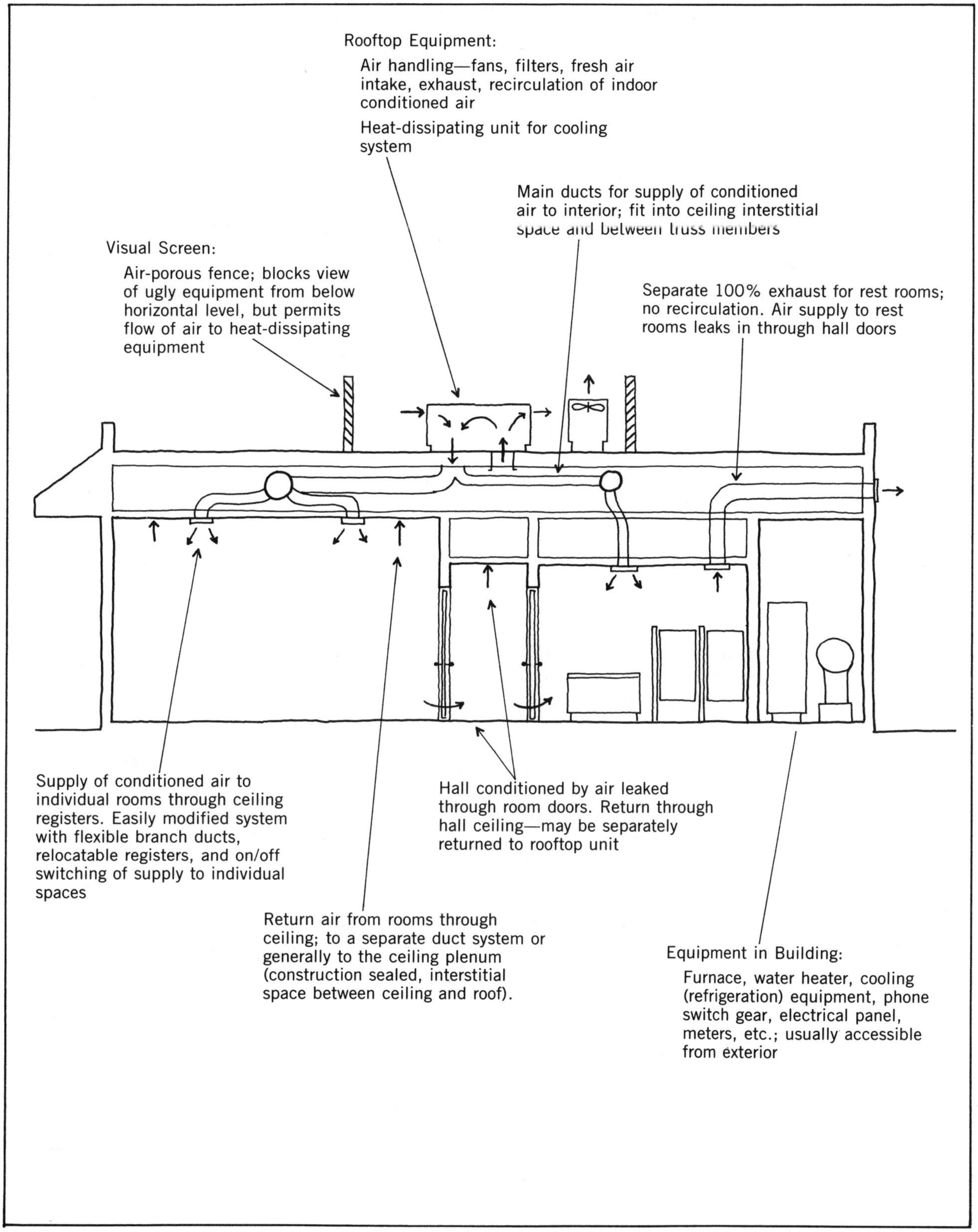

Figure 5.11 Building 2: general scheme and details for the HVAC system.

system normally uses a heat-dissipating unit that must be exposed to outdoor air, requiring it to be placed outside the building—on the roof or at ground level.

The selection of the basic HVAC system, the individual elements of the system, and the general operation of the system must respond to many concerns, principal of which are the following:

The local climate, concerning the range of temperatures, predominant concerns for heating or cooling, prevailing humidity, and wind conditions.

Planning concerns for the building, regarding the type of occupants, need for many individual controls (single interior spaces), concerns for noise and vibration of equipment, and the need for concealing of equipment—both inside and outside

Specific form of construction, concerning particularly provision of wall and/or ceiling interstitial space, ease of access of hidden elements for maintenance or alteration, and natural thermal responses of the construction

Desired locations for equipment and for ducts, registers, controls, and so on

For the system as shown here, some specific concerns that relate to the construction are the following:

Support of the rooftop equipment and separation or isolation for control of vibration and noise. This becomes more critical when the roof structure is longspan and when the equipment is directly above noise-sensitive interior spaces.

Accommodation of the ducts, registers, and return grilles in the ceiling construction and the roof/ceiling interstitial space. This may critically affect the choice of the whole roof and ceiling construction systems.

Accommodation of the heating and chilling equipment in the building interior, as a space assignment concern and a noise separation issue.

Installation details for the piping of the hot and chilled water and for wiring for electrical power and electrical signal control systems.

Use of the light wood frame with its hollow stud walls and the trussed roof with suspended ceiling results in considerable potential for accommodation of hidden elements of the HVAC system. However, this construction does not provide much help for passive thermal response or noise and vibration separation, requiring some considerable enhancement where these are major concerns.

Figure 5.12 shows the construction at the rest rooms. The wall between the rest rooms and the adjacent office is enhanced for sound separation. Some of the details for accommodation of the plumbing are also shown. Note the use of the thickened wall where wall-mounted fixtures are placed on opposite sides of the same wall.

Lighting for this building will most likely be provided by four separate, although interactive, systems. These are:

A daylight-using system, using light from windows, glazed exterior doors, and—possibly—from some skylights or roof windows.

Permanent, hard-wired fixtures, mostly for entry areas, exterior lighting, rest rooms, equipment rooms, and any permanent interior corridors.

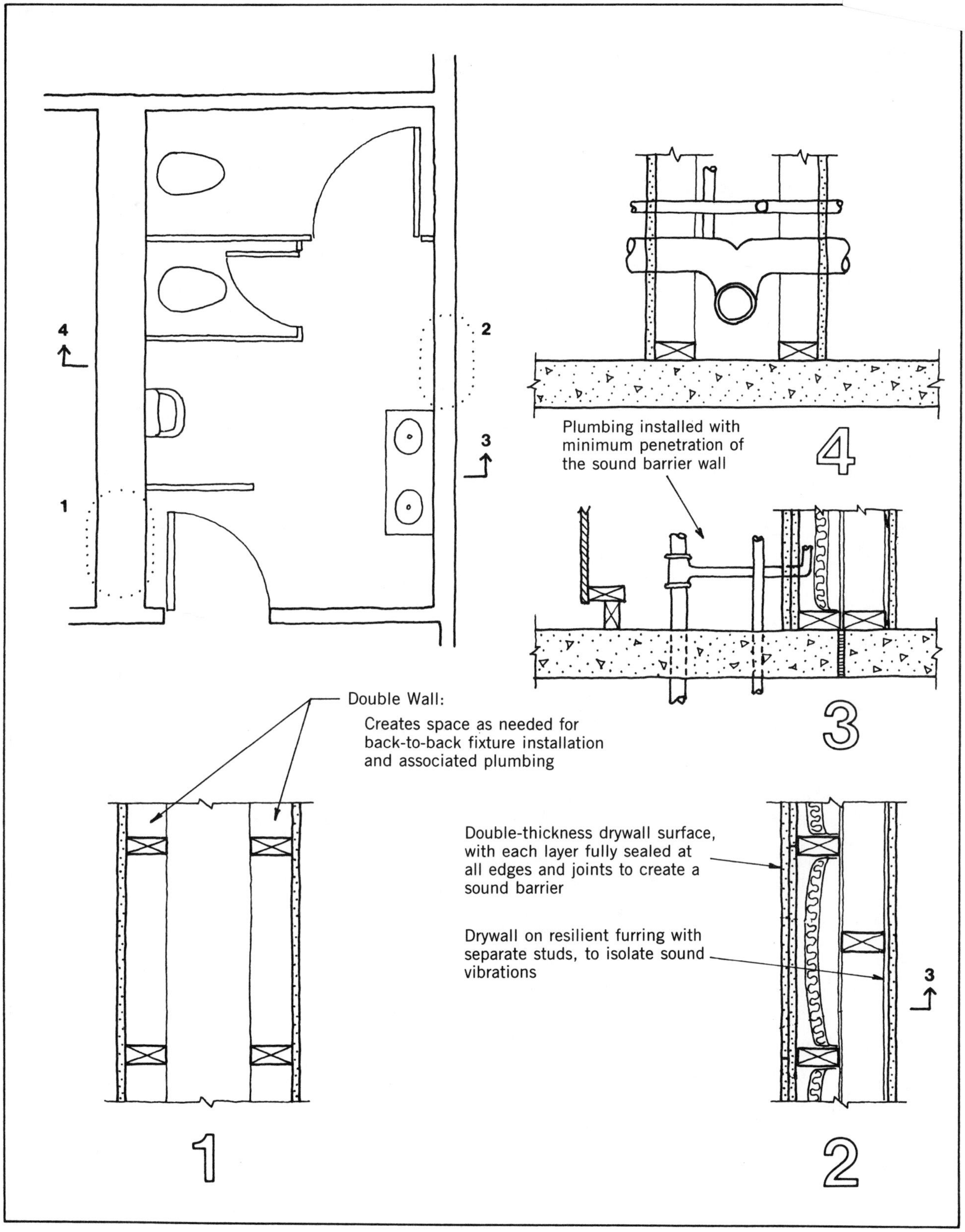

Figure 5.12 Building 2: plumbing systems and details.

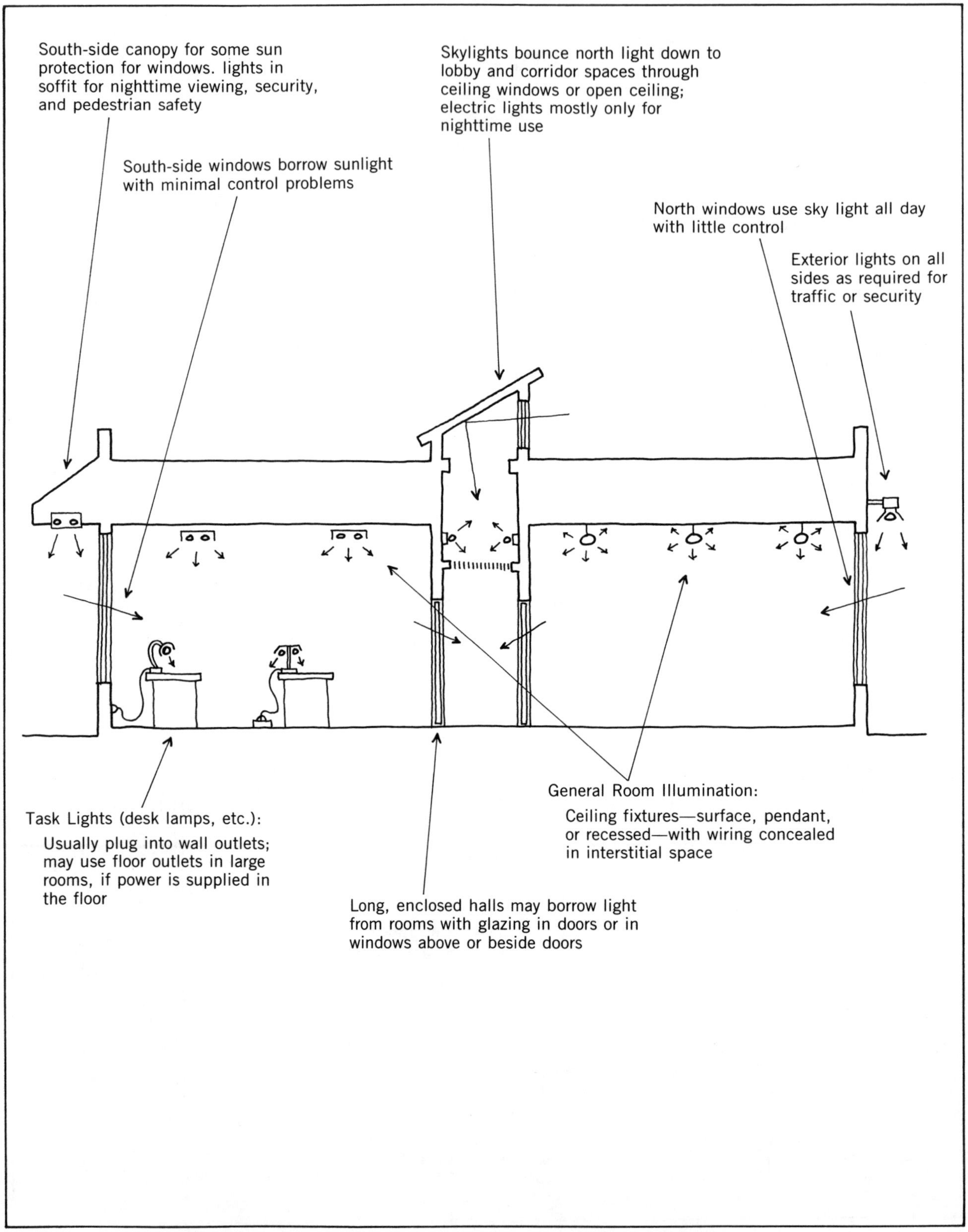

Figure 5.13 Building 2: details of the electrical lighting systems.

A semipermanent, general illumination system, typically installed in response to needs of individual tenants or as a component of a general interior design development of the ceiling or other user-responsive interior construction. This system must respond to the same code requirements as the permanent elements, but is usually subject to modification.

A serendipitous, expedient, instantly changeable, user-installed system, using movable units (floor lamps, desk lamps, etc.) and plugged into the general electrical outlet system.

The designers of the original, and basically permanent, building construction have control over the choices and details for these systems more or less in descending order of their listing above. Windows and skylights are essentially permanent; desk lamps and floor lamps are almost totally out of range of control. Still, the use of the whole array of lighting elements must be anticipated.

Some details for the permanent and semipermanent electrical lighting systems are shown in Figure 5.13. As with the HVAC system, the construction here is quite accommodating for installation of various forms of ordinary fixtures. For the movable, user-installed lighting—as well as for other electrical power needs—some electrical outlets must be provided in the interior construction, notably the interior partition walls. As these walls can only be partly anticipated in terms of location and form, some thought must be given to the means for installation of wiring, outlets, lighting switches, and so on. These matters are discussed more extensively in other building examples in this chapter. Here, the most likely means for wiring of interior, movable partitions would be from the ceiling down through the hollow walls.

Figure 5.14 shows some details for the installation of a sprinkler system and a smoke alarm system. Need for these systems and choices for these or other fire-fighting and alarm systems may respond to local codes or simply to building owner or tenant requests. This construction is obviously highly combustible and fast-burning, so any help in this area will greatly enhance fire safety in general. However, fire hazard is often more critically present in the form of furnishings, finish materials, and other building contents that are often largely out of range of control by the building designers. Major efforts to make the basic construction highly fire-resistive may provide a false sense of security for occupants.

Figure 5.15 shows the general site plan and some details for the installation of a sewer system for storm water (rain) drainage from the roof and general lot spaces. Site surface drainage must relate to general conditions regarding neighboring streets and properties. Much of the runoff from paved areas near the site edges will likely feed into curb gutters and sewers on public land, so this is a highly constrained factor for site development.

For this quite flat site, the large parking area at the front of the building requires some area drains, which feed into the same sewer lines as the roof drains. The plan shows this sewer line feeding to an undesignated sewer main beneath the frontage street.

In addition to the buried storm sewer shown in Figure 5.15, there are likely to be buried lines for:

A sanitary sewer for wastes from the rest rooms

Wiring for the building electrical power supply

Water supply piping

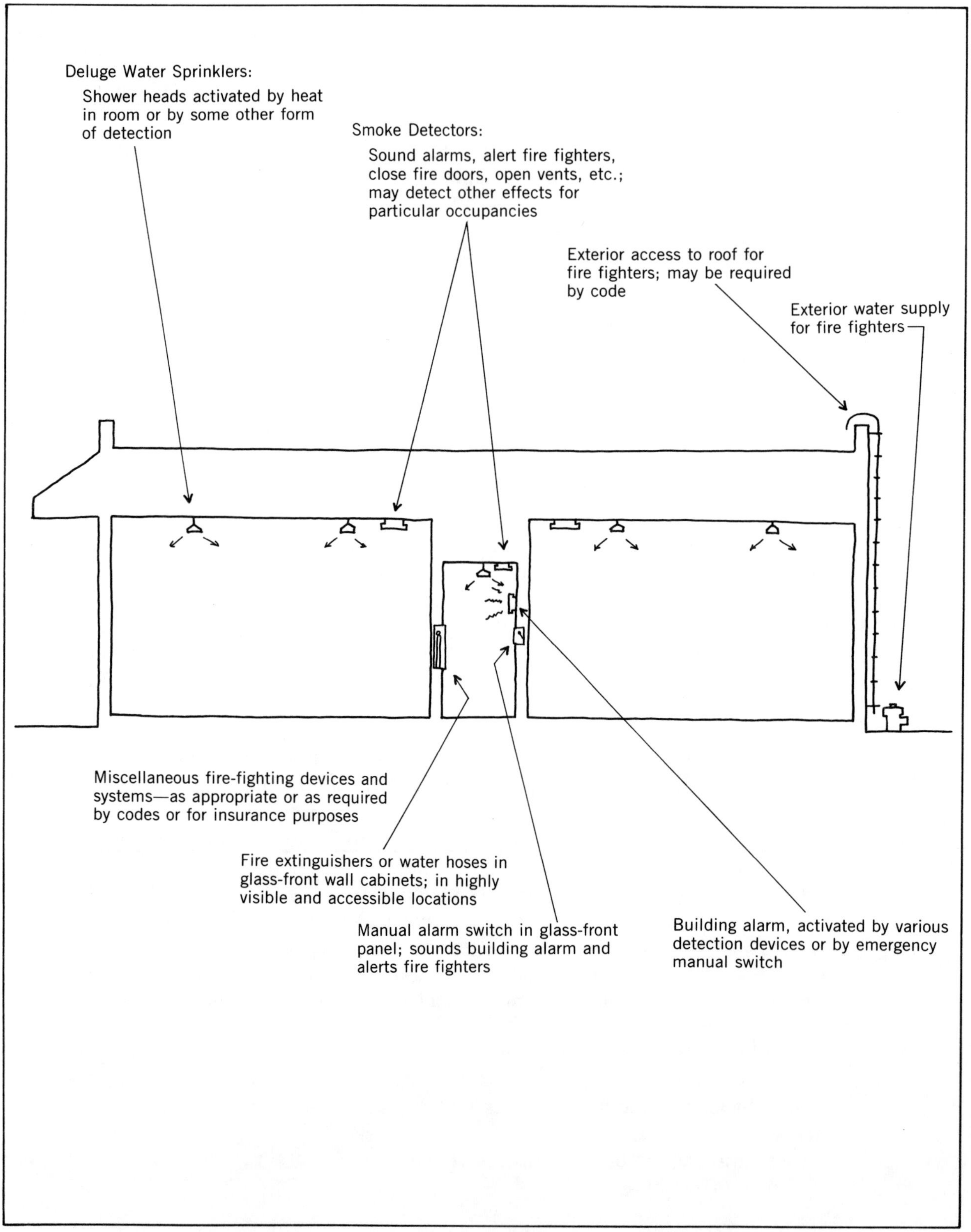

Figure 5.14 Building 2: details of the fire-fighting systems.

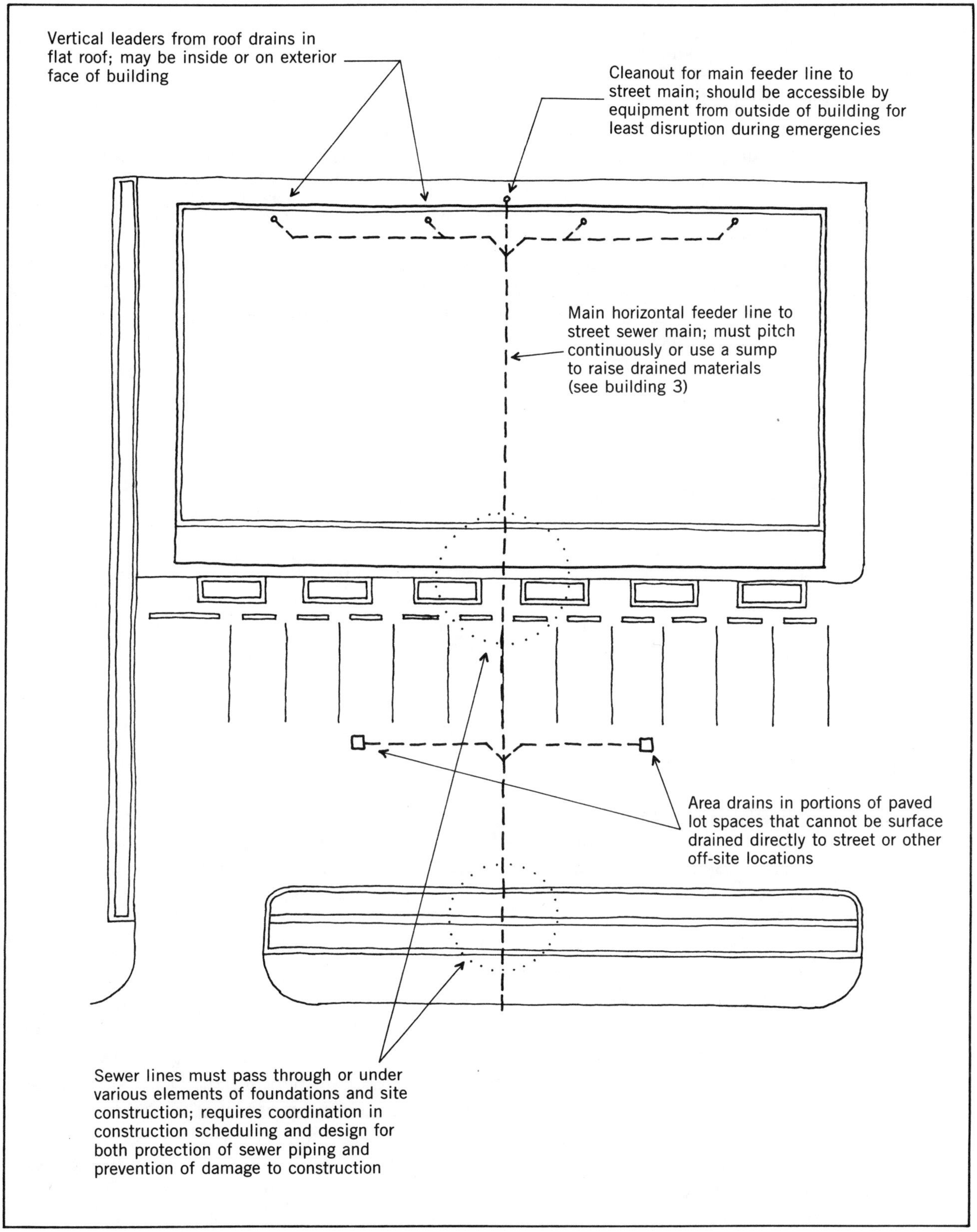

Figure 5.15 Building 2: details of the storm water drainage system.

Wiring for telephones and possibly for cable TV

Gas supply piping

Wiring may alternatively be supplied overhead (above ground, from utility poles), but the piped systems will usually be underground.

Planning, detailing, and scheduling of the site and building foundation construction must be done with full knowledge of the needs for these buried systems. This is generally routine stuff, but it must still be accounted for in the building, site, and landscape designs.

An important detail concern for the sewer lines is the provision of cleanout access points for the all-too-frequent possibility for clogging of the systems. If these are not well coordinated as to location and details for access, the servicing of the systems can be very disruptive.

5.3 BUILDING 3

One-Story Industrial Concrete/Masonry Walls, Frame Roof Large Flat Site

This is a multiple-bayed building, indefinitely extendable in two directions, producing a large covered floor space for an industrial plant, warehouse, or other such usage (see Figure 5.16). Depending on requirements for clear spans, wall heights, clear height beneath the roof structure, and total floor area, various structural elements may satisfy code requirements, user needs, and the general desire for very economical construction.

As shown here, the roof has a slow-draining, almost flat profile. Drainage of such a roof is a major problem as areas near the center of the large building are some distance from a building edge. Edge draining, as shown for building 2, is generally not feasible here, as the total amount of the required pitch from the center to an edge would be prohibitive. Some edge draining may be provided, but some roof drains of the type used for interior drainage will be required, most likely discharging into interior vertical leaders located at columns and in turn feeding into an underfloor piped sewer system.

Roof slopes will most likely be achieved here by simply tilting the elements of the roof framing. The nonflat bottom of such a frame is not a concern if no ceiling is provided.

Choice of exterior wall materials will depend on fire requirements, local climate, interior environmental control requirements, need for a particular type of window or for many large windows, concerns for exterior appearance, and other possible considerations. Typically, cost will be a major factor, so the most economical options will quite likely be preferred.

A critical factor for the choice of the wall construction will be the height of the wall. These buildings often require considerable interior height; the walls thus become quite tall and must usually span vertically for lateral wind pressure or seismic thrust. The walls may be developed as framed structures with columns and curtain walls, or with a major structural form of wall construction.

The two schemes shown in Figure 5.17 use structural walls with potential for use as both bearing and shear walls. In mild climates, the wall construction shown here may

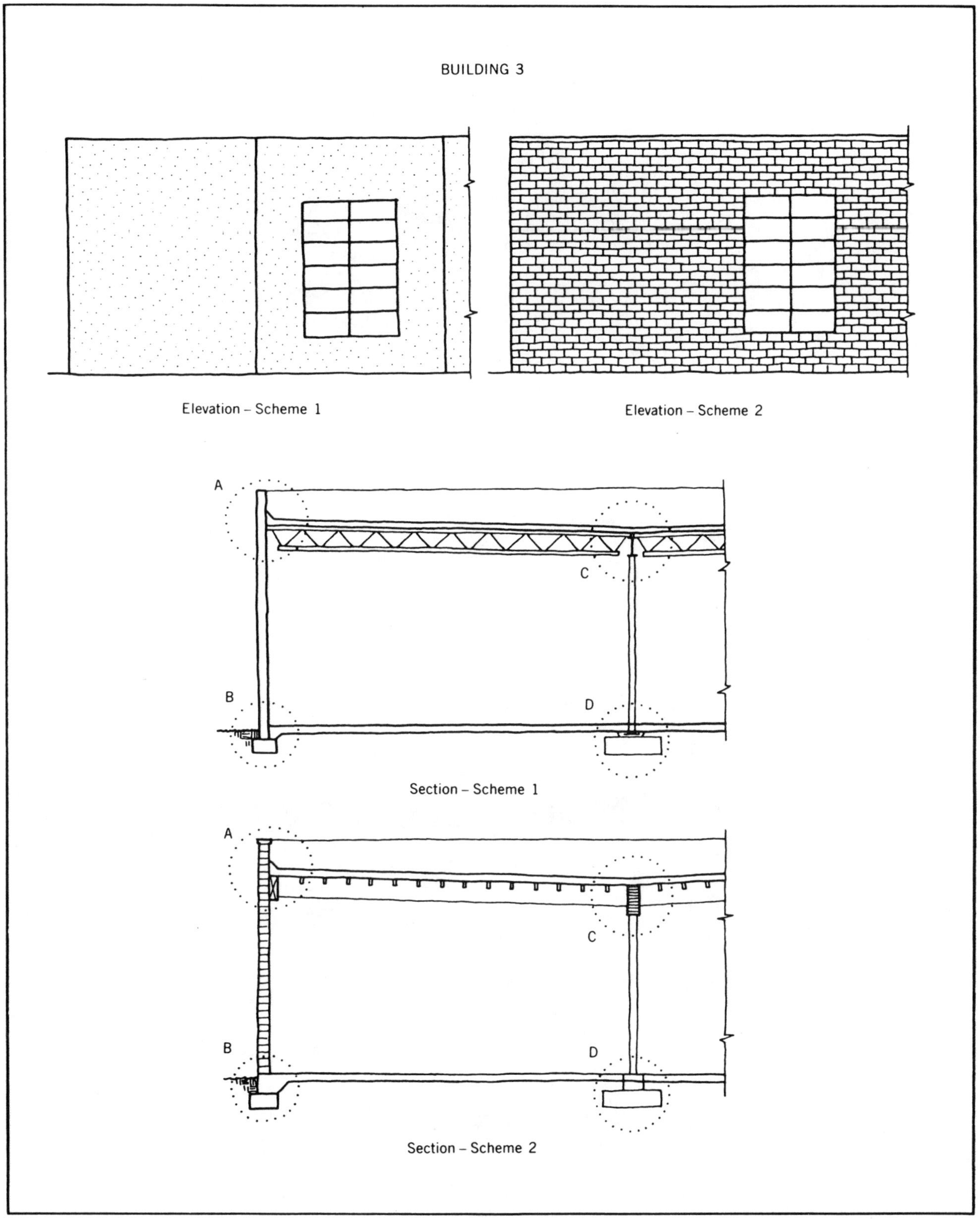

Figure 5.16 Building 3.

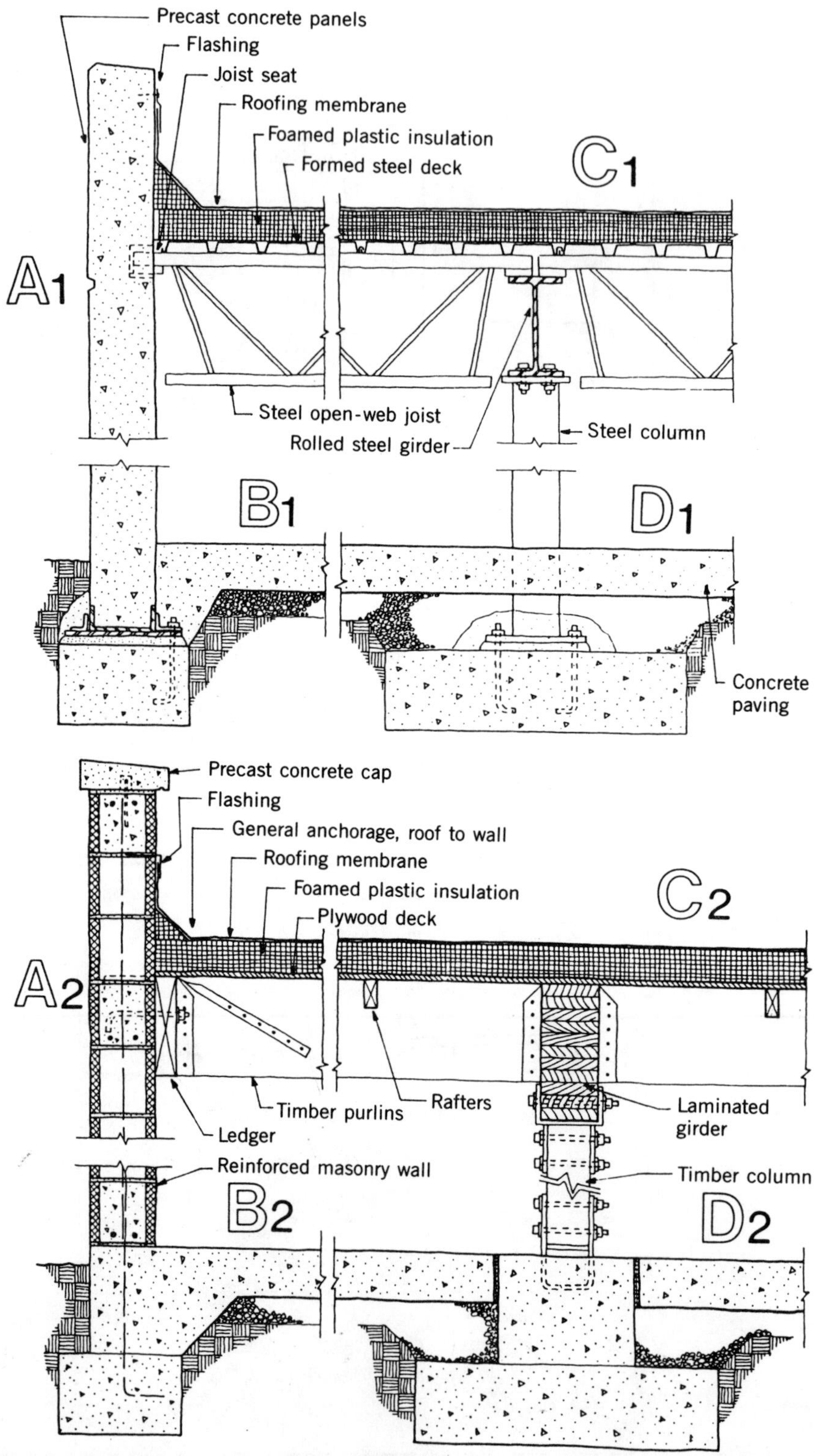

Figure 5.17 Building 3: details of the building shell.

be used with no additional materials. In cold climates, however, some enhancement with insulation and moisture barriers will be required, making an insulated sandwich panel system possibly a better choice.

Options for the roof structure are considerable in number, and the most economical system will probably be chosen. The two systems shown here are common and can be used for a range of spans.

Floor slabs are usually left uncovered, although surfaces may be treated in some manner to produce a better than usual hard, smooth face for wear resistance. Any floor treatment is possible, but rough use and economics usually dictate the use of the exposed concrete surface.

While the general interior of such a building may be as shown in Figure 5.17, there is usually some need for partitioned spaces within the general enclosure—for toilets, offices, equipment, and so on. Such spaces may be fully enclosed with construction that includes the development of partitions, a suspended ceiling, and some enhancement of the inside surface of the exterior wall. Floor treatments may also be used in such a space, although the change in finished floor surface level will present a minor problem at doorways between the two types of spaces, which probably favors use of relatively thin flooring materials.

Figure 5.18 shows some details for an HVAC system for building 3. This is actually a two-part system, with one system for the general building space and another for the enclosed spaces within the general building envelope. The latter will most likely be a relatively conventional system, similar in general to the options for building 2. Supply and return for this system can be incorporated into the generous space above the ceilings in the enclosed interior spaces. Depending on the building plan and the nature of the systems, the two systems for this building may be totally separated with little or no sharing of equipment or housing space within the building.

The system used for the large, general building space will very much depend on what happens in this space. Special activities involving spray painting, sawdust, toxic fumes, and so on, may require very special exhaust systems and particular details for general air circulation. Use of very sensitive equipment or storage for special materials may establish a narrow range of tolerance for changes in temperature or humidity, even narrower than that for ordinary human comfort. For ordinary situations, however, this system may be similar in general form to that for building 2, merely larger in all respects—with bigger ducts, fans, and other equipment elements, more registers and returns, and so on.

Instead of the single, centrally operating and centrally controlled system as shown in Figure 5.18, it may be possible to use a series of individual, rooftop systems, each serving a limited area of the building. This is more likely in some climates and when different needs occur at different locations in the building.

In general, the accommodation of the equipment and various elements of the HVAC system is usually somewhat easier in this type of building. Appearance and the general need to conceal elements from view may be of little concern. Vibration and noise of equipment may also be of less concern, unless they actually disturb some activities in the building. While the masonry or concrete exterior walls provide no basic interstitial spaces, the simple surface mounting of wiring or piping is probably of little concern. All in all, this is typically a very barren form of construction, with simple functionality of prime concern—which will most likely extend to the simple needs of the various service systems.

As mentioned previously, a typical problem for large buildings of this type is the need for some interior roof drains. Some details for such a system are shown in

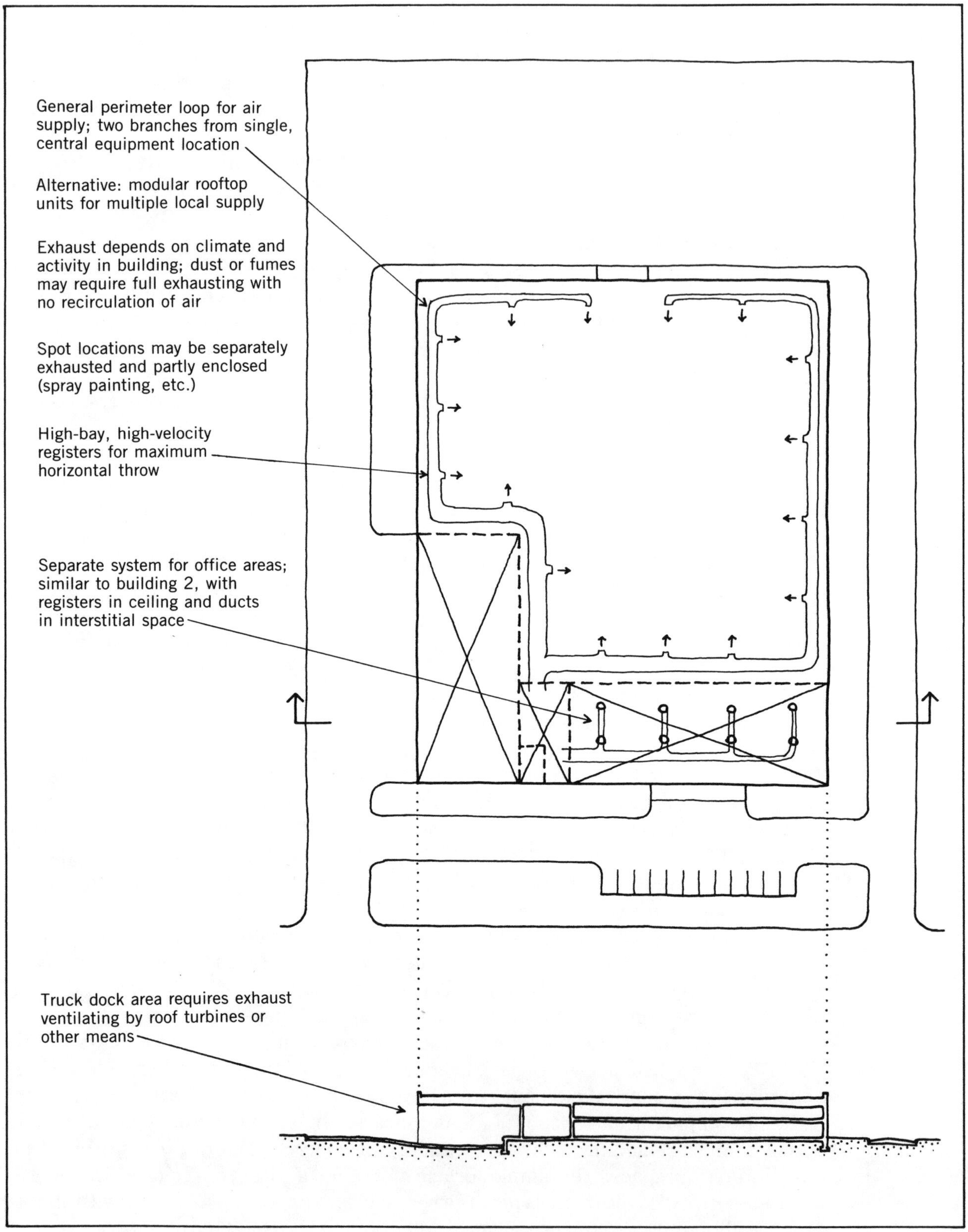

Figure 5.18 Building 3: general scheme for the HVAC systems.

Figure 5.19. The roof drains will be sized for a specific area of roof and its runoff based on local weather. Location of the drains must be at low points of the multiple-sloped roof surface, a pattern for which must relate to the general roof construction and the location of major roof structural elements.

Although roof drains may be located anywhere in the building plan, the vertical leaders from them are usually located at interior columns. The leaders must feed into some underfloor sewer systems, possibly involving some routing of the piping to avoid conflict with the column foundations, as shown in Figure 5.19.

The underfloor sewer system must eventually leave the building and make a run to some off-site storm sewer main. For the large building—possibly on a large lot—this may involve a considerable horizontal distance. Since the sewer must work by gravity flow, this can put the line at considerable depth by the time it reaches the sewer main. A device sometimes necessary to handle this problem is a sump pit with a submersible pump, which is used to raise the drained water to a higher level. The general form for such an installation is shown in Figure 5.19, which shows the pit inside the building and the discharge from the building raised as high as possible to keep the line below the site surface at the building edge.

Sump pits must be accessible for service, both to remove accumulations of sediment and to maintain the pump. This is probably better achieved with the pit outside the building where site conditions permit.

5.4 BUILDING 4

Two-Story Motel
Masonry and Precast Concrete

As shown in Figure 5.20, this building has a common form of plan for motels, with rooms off a central corridor. The same basic structure might also be used for a dormitory or small apartment building.

The basic structure, as shown in Figure 5.21, consists of structural masonry walls and precast concrete roof and floor units. Although this is a widely used construction, there are many possibilities for variations, both in the general structure and in finish materials and details. Higher quality (supporting higher room charges) is usually achieved with better finish materials, although some quality may be achieved by enhancements of the basic construction as well.

Choices for the masonry wall structure depend largely on regional considerations, involving both climate and type of critical lateral force effects. The exterior wall construction shown here utilizes reinforced concrete block (CMUs) as the structure with an exterior brick veneer. This is a system favored in areas with severe windstorms or high seismic risk, and the general form of the structure would change in other regions where seismic response is not critical.

In cold climates the masonry walls need considerable enhancement to improve the character of thermal flow. The noninsulated walls shown here would be possible only in a very mild climate. Insulation can be developed in a variety of forms, including the following:

Foam plastic units adhered to the inside surface of the wall with gypsum drywall adhered to the insulation units

Glass fiber batt insulation in a void space created by furring out on the inside of the masonry, with gypsum drywall fastened to the furring strips

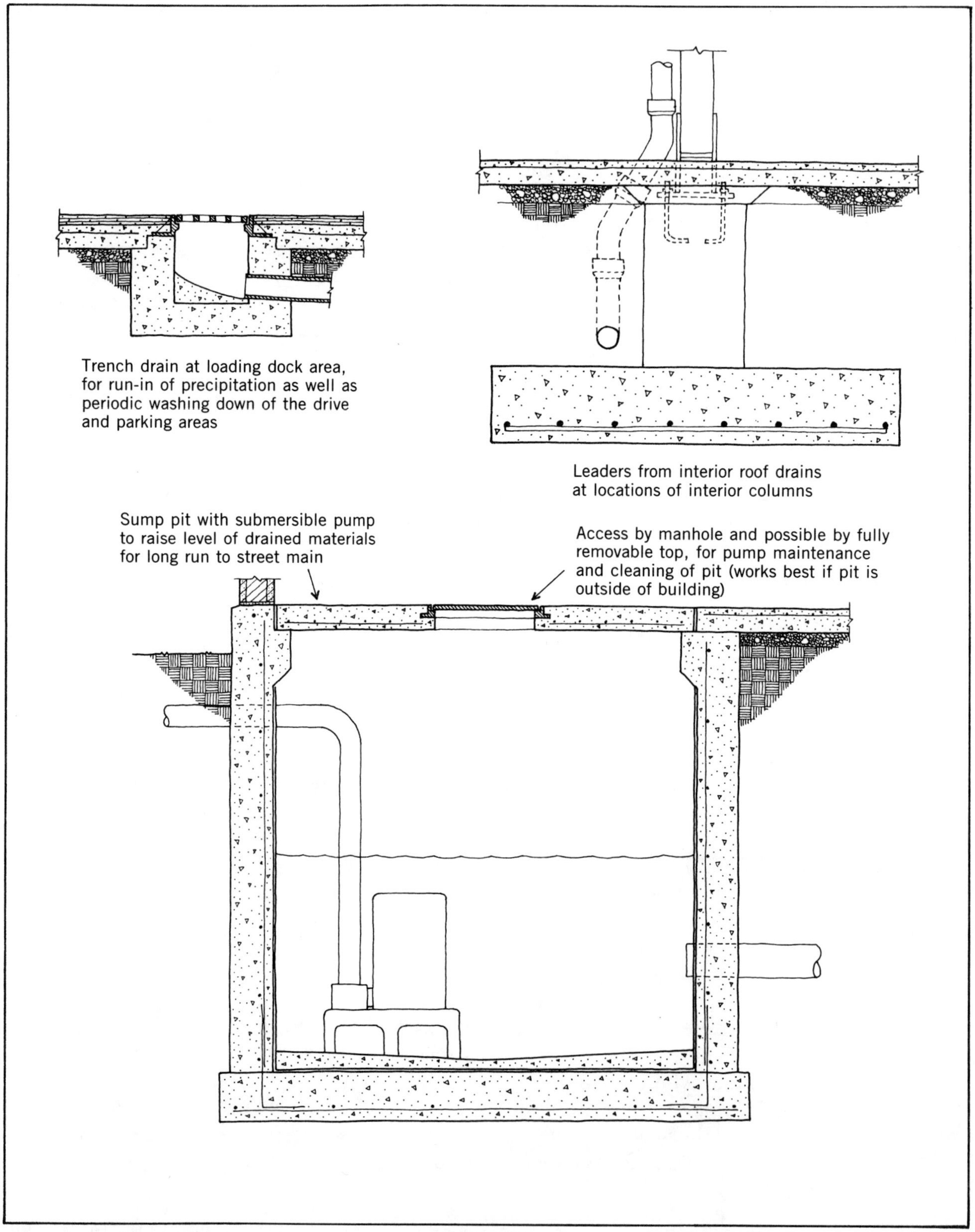

Figure 5.19 Building 3: details of the interior roof drain and storm water sewer system.

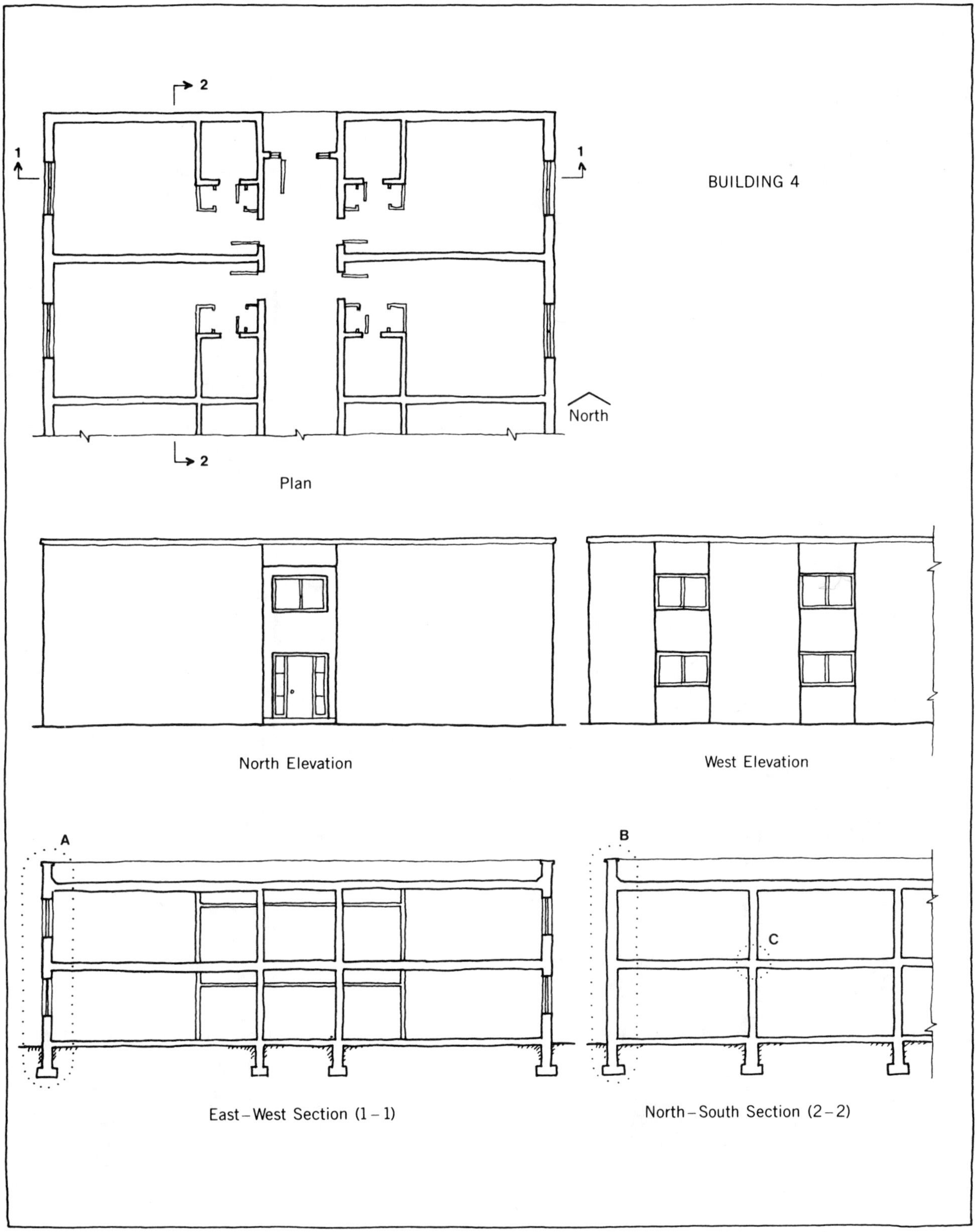

Figure 5.20 Building 4.

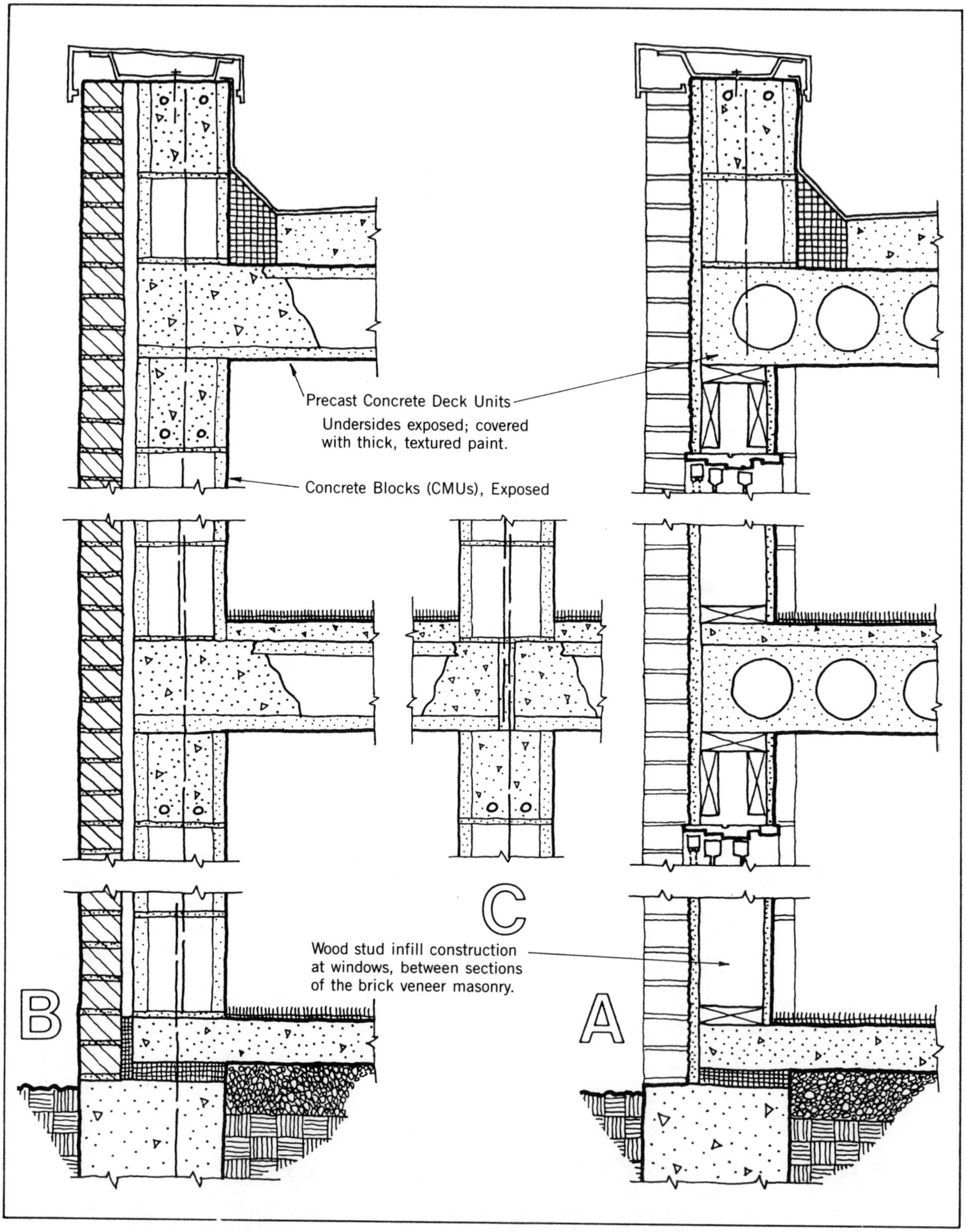

Figure 5.21 Building 4: details of the building shell.

Foam plastic insulation in the cavity space between the CMUs and the brick veneer

Foam plastic units adhered to the outside face of the masonry (without the brick veneer) with a priority vinyl stuccolike finish with fiberglass mesh reinforcement

The development of interior wall finishes must first relate to any considerations for the exterior walls, including those just described. The walls at the windows are developed as light wood frame, with all of the usual options for surfacing on the inside of the frame.

Development of interior masonry walls is shown in Figure 5.21. The walls between motel rooms are structural CMU construction, providing support for the roof and floor units. These are relatively heavy walls and may well be left exposed, possibly receiving only some paint. However, for a higher-quality finish, they may receive some surfacing with paneling or plaster. Their sound privacy quality may also be enhanced by using furred-out surface on one side, with an additional degree achieved by using resilient attachment of the surfacing to the furring.

The nonstructural partitions for the halls, bathrooms, and closets would most likely be achieved with simple light wood framing. A dropped ceiling is typically used in these areas to allow for encasing of air ducts, wiring, and piping. This form of construction is shown in the details in Figure 5.22.

Figure 5.23 shows some details for the HVAC and plumbing systems for the motel, with elements commonly used for low-rise buildings. Each motel rental room must be provided with a bathroom and have some individual control for ventilation and thermal conditions. For the double-loaded corridor plan, there is also a need to ventilate the corridors and interior stairs.

Shown here is the use of a through-the-wall unit for each room that combines the functions of ventilating fan, air conditioner (cooler), and heater. In a typical configuration, this unit is placed beneath the window in the exterior wall. A separate, centrally operated air exhaust system draws air from the bathroom, typically operating continuously to assure some minimal continuing air exchange in the rooms.

Depending on climate conditions, corridors and stairs may have no direct HVAC control—with some reasonable condition maintained by leakage from rooms and from the outdoors—or have a separate system for ventilation and/or thermal control. Details of the interior construction in Figure 5.23 show the common use of a furred-down ceiling space over the corridor and the bathroom and entry areas inside the rooms. This space is used for housing of air ducts and some of the piping for the bathrooms. Use of the dropped ceiling is essential here in conjunction with the precast concrete floor and roof systems, which provide for little useful interstitial space. If a nonexposed wood or steel framed system were used, the dropped ceiling might not be necessary.

Figure 5.24 shows some details for the installation of the central elevator system, indicating the use of a hydraulic elevator—a common solution for the low-rise building.

5.5 BUILDING 5

Church Auditorium/Chapel Exposed Timber Structure

This is a single-space, medium-span building with essentially no interior structure. The options possible for this building in terms of type of structure and architectural

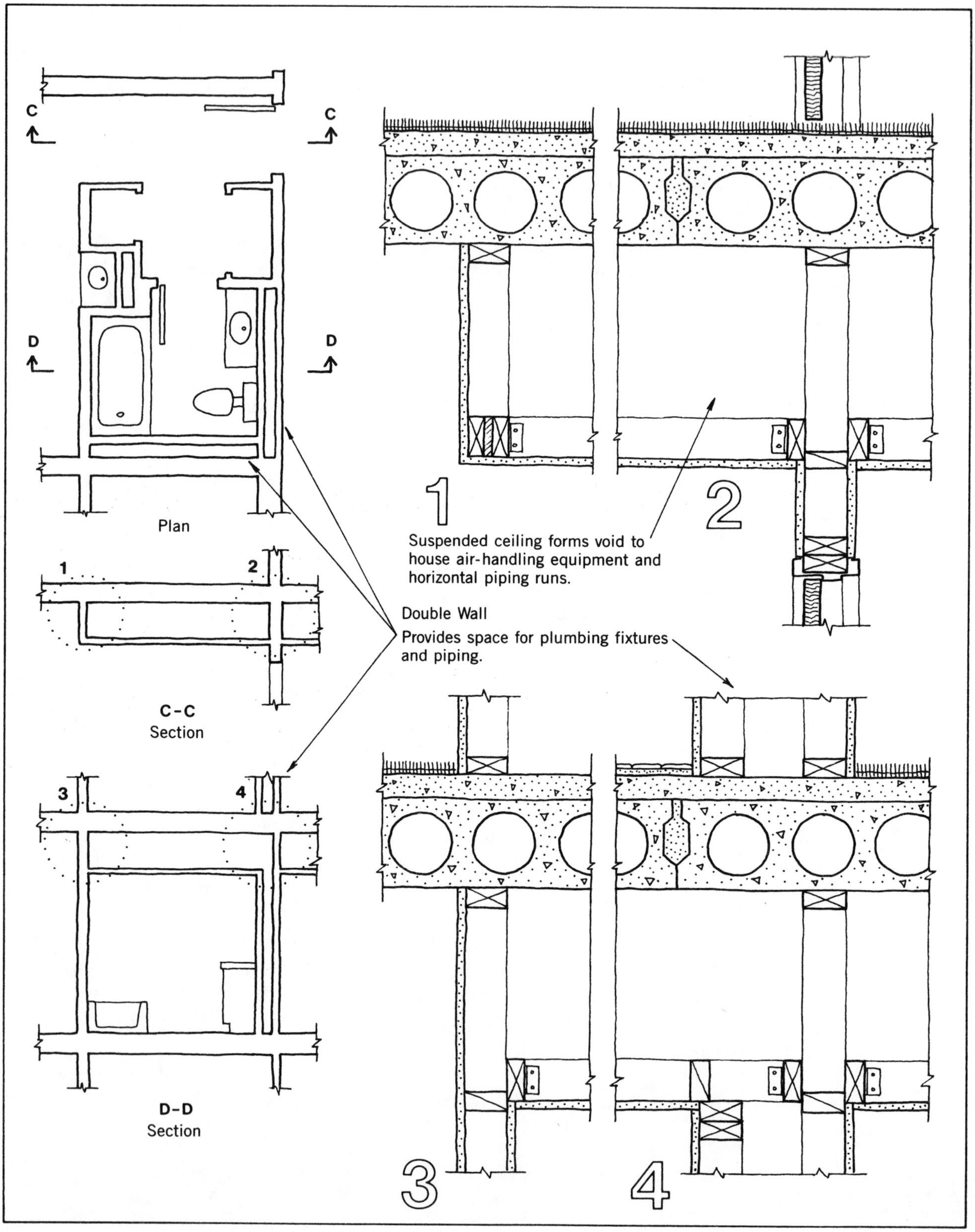

Figure 5.22 Building 4: details of the interior construction.

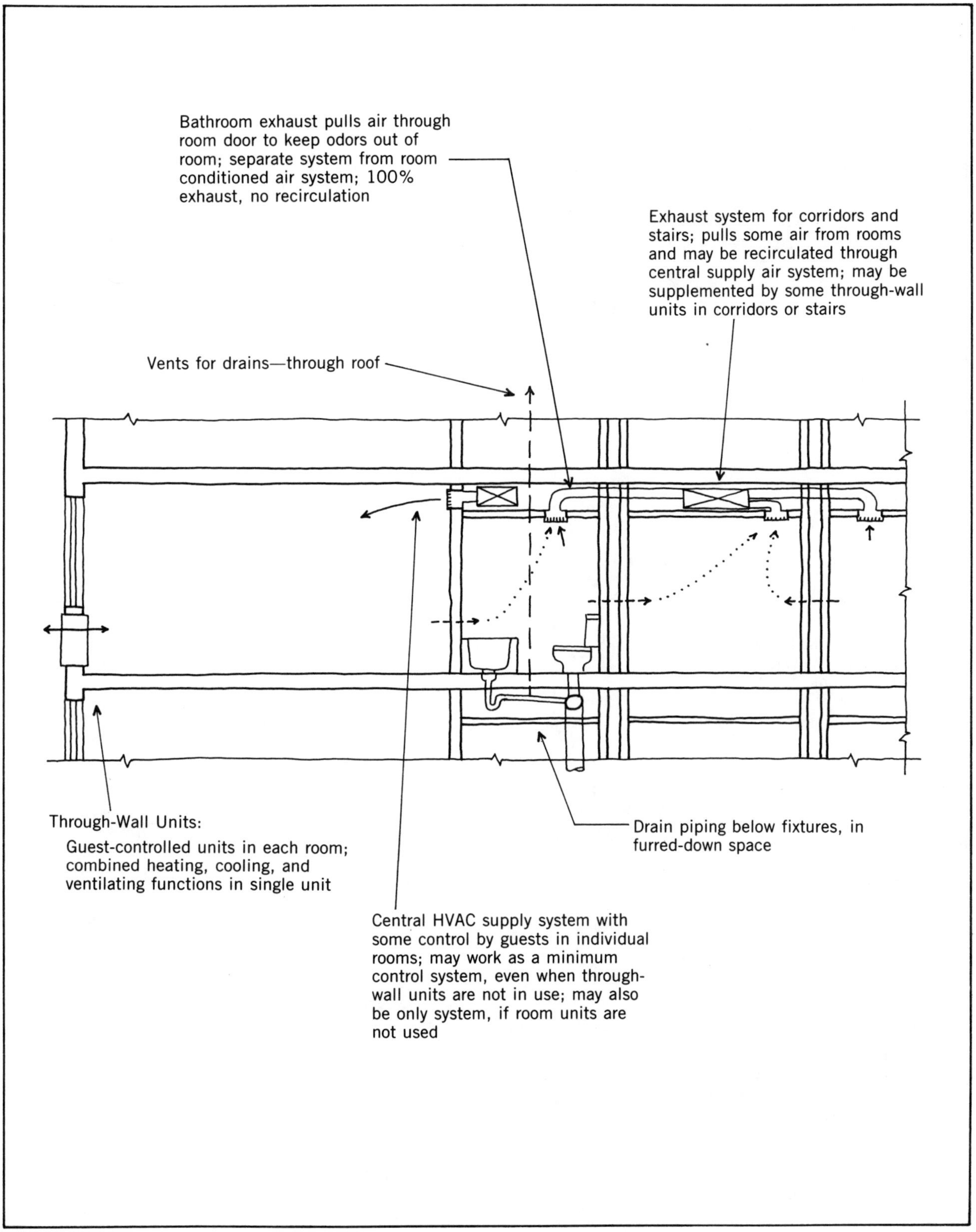

Figure 5.23 Building 4: details of the HVAC and plumbing systems.

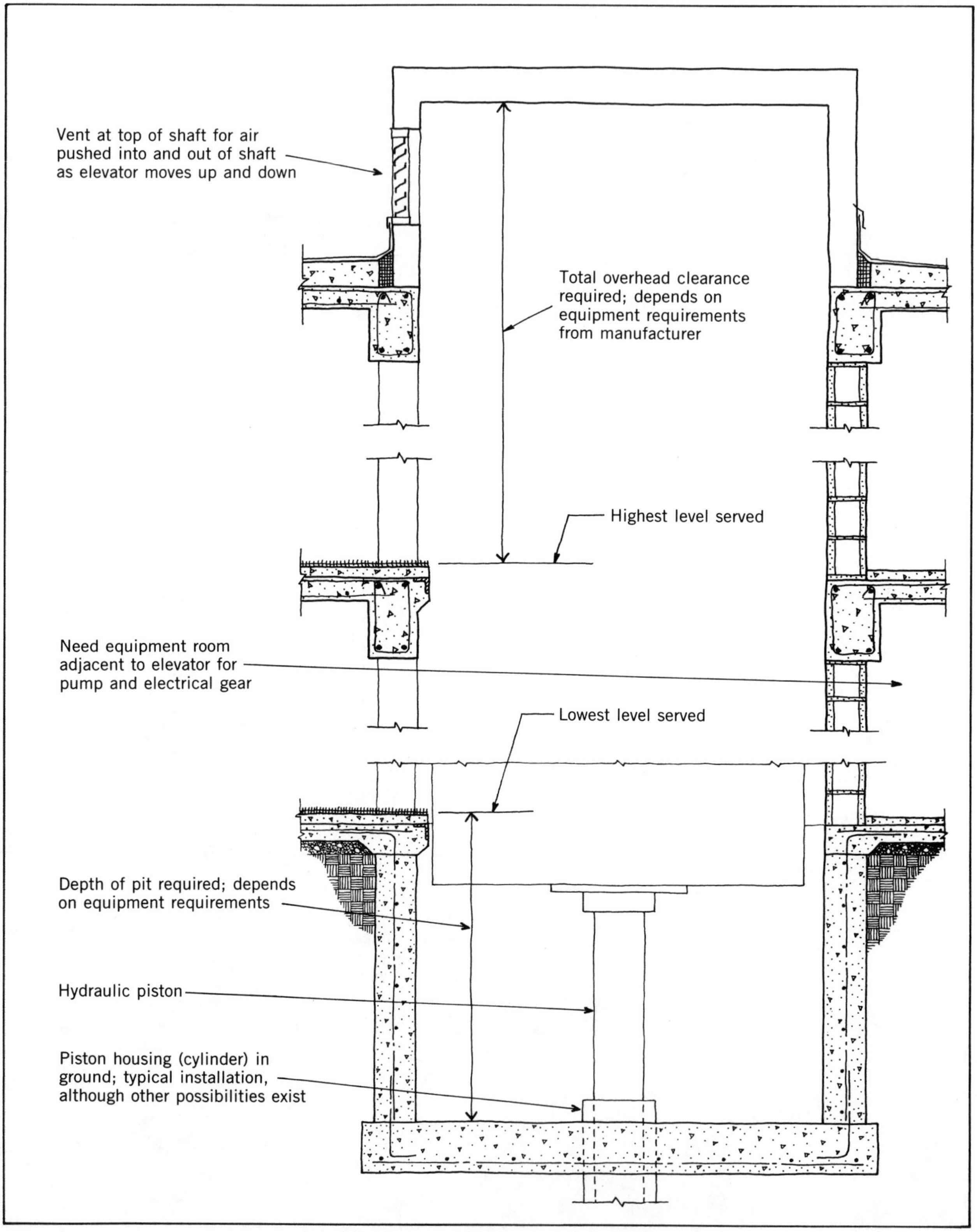

Figure 5.24 Building 4: elevator details.

style are virtually endless. Shown here is a popular form of construction, with gabled bents of glued laminated wood and an exposed wood beam and plank deck roof. Infill walls between the gabled bents are achieved with light wood frame construction.

The general form of the building and the vaulted interior are shown in Figure 5.25. The front podium/altar area is slightly raised as a stepped platform. The other end of the church may well contain some rooms and a balcony with an entry area and stairs, but we will deal only with the auditorium portion of the building shown here.

The basic construction of the building shell is shown in the details in Figure 5.26. The building interior is comprised primarily of the insides of the roof and exterior walls and the top of the concrete pavement. Further development of the interior consists mostly of choices for the finishes on these elements.

With the exposed form of construction for the roof, there is no suspended ceiling or other form of interstitial space in the upper portion of the auditorium. This poses some severe limitations on the systems for HVAC and lighting, especially if general concealment of the installation is desired. A general overhead ducted HVAC system is out of the question, unless the whole thing can be hung out for display. This is possible, of course (see Figures 2.3 and 3.2), but does not fit well with this interior and general construction.

Other options for the HVAC system for the auditorium depend somewhat on local climate and the overall use of the building. Churches are often used only at certain times, so a system may be able to be shut down—or at least operate at limited capacity—for extended periods. However, when the space is to be used, it is usually desired to bring it to a comfort level fairly quickly, a performance best had from an all-air system. If the space sees a lot of ongoing use, however (for church, school, and general social activities), it needs a system that is similar to that for other buildings with regular usage.

A possibility for an all-air system is shown in Figure 5.27, employing underfloor ducts that feed perimeter floor-level registers. The registers may be in the floor, as shown, or possible in the wall beneath the windows. A variation on this would be to have the registers high in the walls, possibly above the windows, which would work somewhat better for delivering air to the center of the space.

A consideration for very cold climates—here as in other buildings with concrete slab floors—is the possibility for heat loss and very cold floors at the building edges. Various forms of edge insulation may alleviate this, but the use of the perimeter heating system is one possible solution for warming up the floors and the bottoms of the exterior walls.

Another possibility for cold climates is the use of a radiant system in the floor, probably using piped hot water. This could be combined with an air system to provide warm floors and building edges, as well as a minimum general heat level between uses of the auditorium. The air system could then be turned on to quickly raise the temperature for use of the space.

Lighting of the auditorium would probably include a general illumination system and various special lighting for entries, for the the podium area, and so on. The general lighting system may use wall-mounted fixtures with some bounce light from fixtures directed at the ceiling. However, if the ceiling is stained wood, this is less feasible. Figure 5.28 shows a more common solution using pendant fixtures suspended from the underside of the roof, with the wiring concealed in the roof construction.

If the suspended light fixtures are supported from the laminated bents, the wiring could be installed in channels cut into the top of the bents and drilled vertical holes to

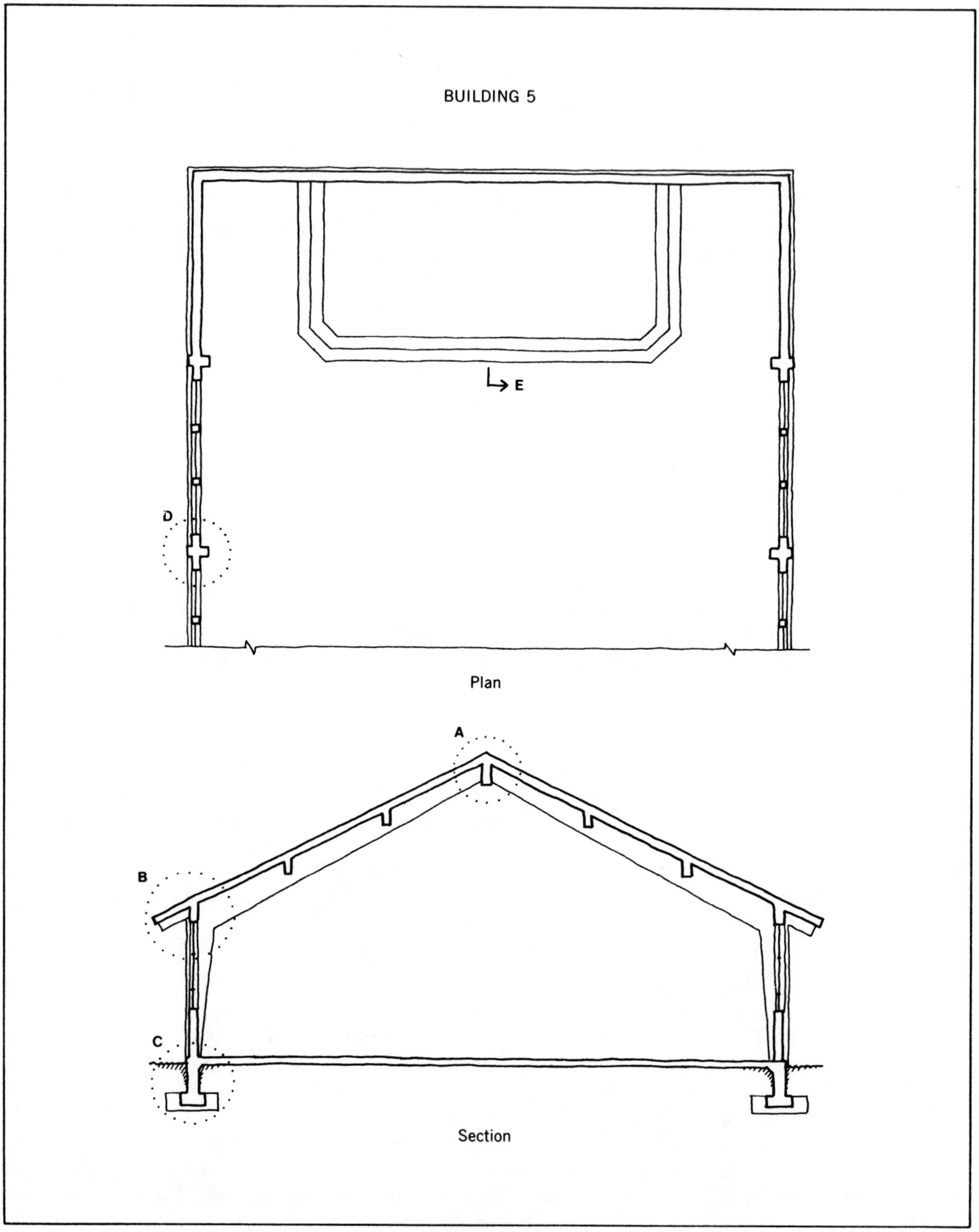

Figure 5.25 Building 5.

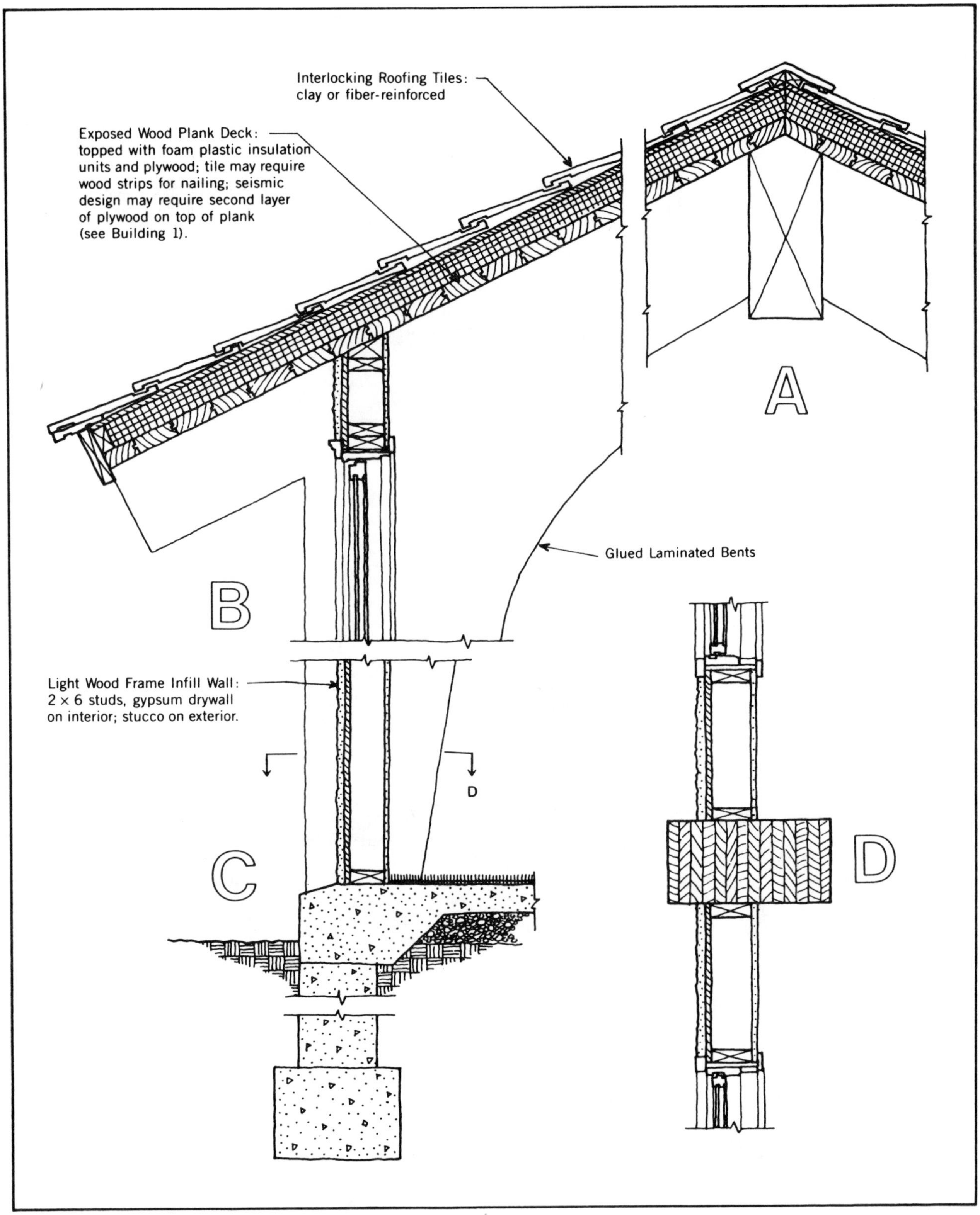

Figure 5.26 Building 5: details of the building shell and structure.

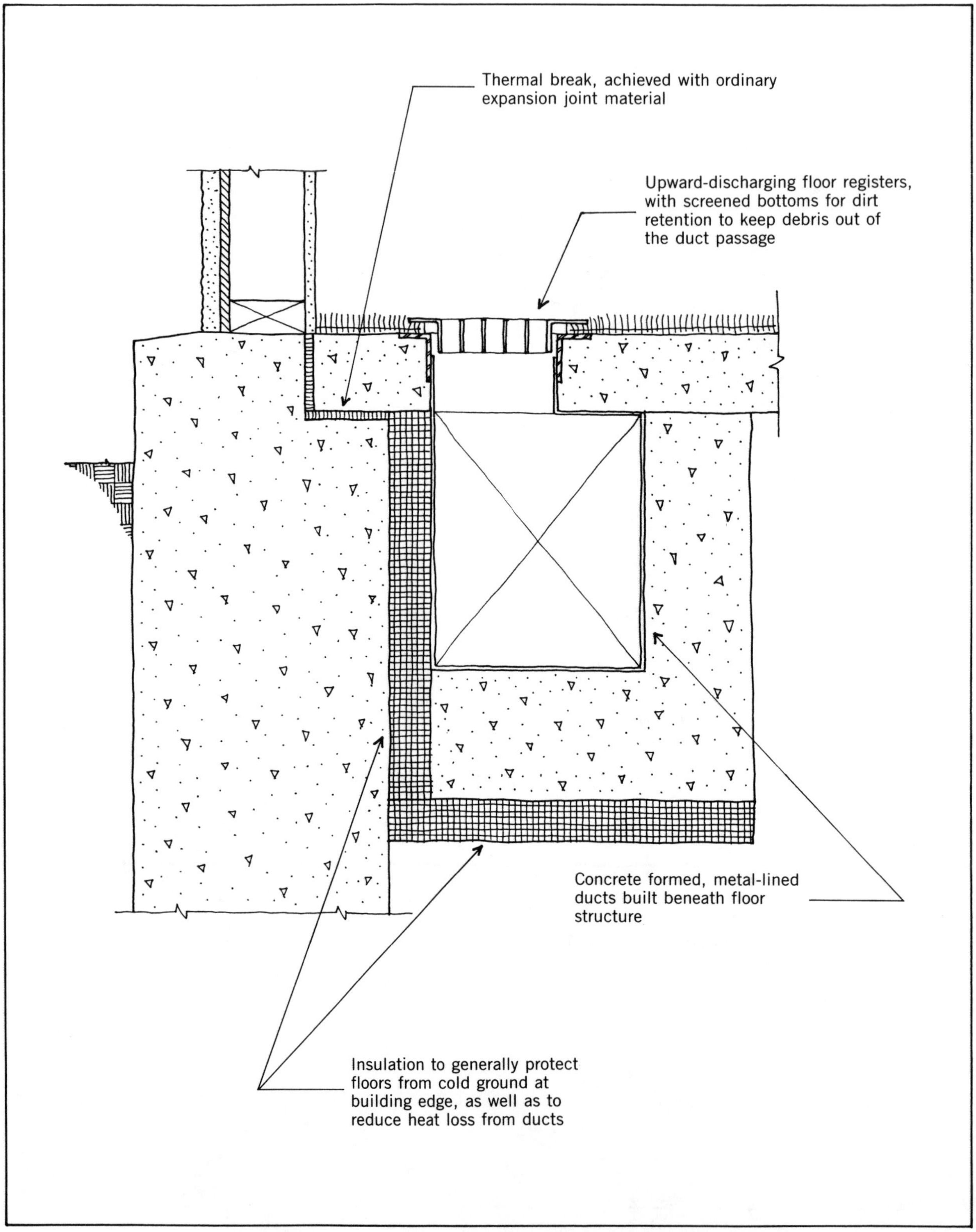

Figure 5.27 Building 5: details for the perimeter HVAC system.

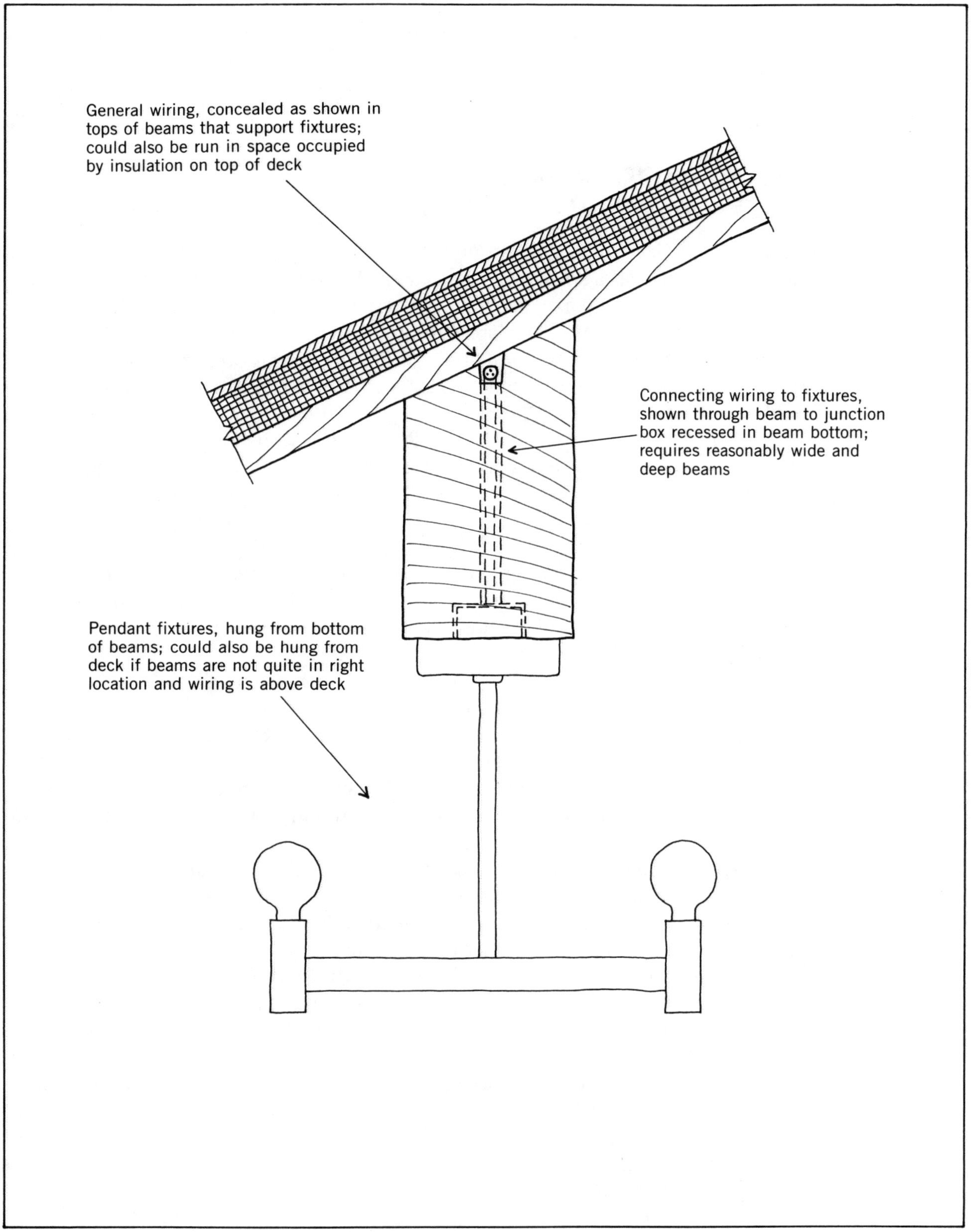

Figure 5.28 Building 5: details for the general lighting system.

the individual fixtures. The bents would have to be oversized for this cutting and the cutting work carefully supervised to preserve the integrity of the exposed construction.

5.6 BUILDING 6

Three-Story Office Building Steel Frame, Metal Curtain Wall

This is a modest-size building, generally qualified as being low rise (see Figure 5.29). In this category there is a considerable range of choice for the construction, although in a particular place, at a particular time, a few popular forms of construction tend to dominate the field. Currently available products, market competition, current code requirements, popular architectural styles, and the general preferences of builders, craft people, developers, and designers usually combine to favor a few basic forms of general construction.

Shown here is a steel frame structure with W sections used for columns, girders, and beams and a formed sheet steel deck for the floors and roof. Many variations are possible with this basic system, mostly dealing with choices for the deck system and the members that directly support the deck.

The wall system here is a curtain wall developed as an infill steel stud system. Windows consist of strips between columns, framed into rough openings in the stud wall structure—not unlike the basic method used in buildings 1, 2, 4, and 5. Details of the basic enclosure system and the floor/ceiling system at the level of the office floors are shown in Figure 5.30.

It is common to use some basic modular planning in this type of building. Spacing of the building columns constitutes one large unit in this regard, but ordinarily leaves a wide range for smaller divisions. Modular coordination may also be extended to development of ceiling construction, lighting, ceiling HVAC elements, fire sprinklers, and the systems for access to electric power, phones, and other signal wiring systems. There is no single magic number for an interior modular planning system, and all dimensions between 3 and 5 ft have been used for partitions and window mullions. In the United States, common modular units are 4 ft, 12 in., and 4 in. Most priority ceiling systems and corresponding lighting systems use the 1-ft/2-ft/4-ft system for basic elements. However, 4-ft elements can be spaced on 5-ft centers or otherwise utilized in other planning modules. Selection of a particular manufacturer's system for interior development may establish some criteria for this planning.

For buildings built as investment properties, with speculative occupancies that vary over the life of the building, it is usually desirable to accommodate future redevelopment of the building interior with some ease. For basic construction, this means a design with as few permanent structural elements as possible. At a bare minimum, what is usually required is the construction of the major structure (columns, floors, roof); exterior enclosing walls; and interior walls around stairs, elevators, rest rooms, equipment rooms, and risers for building services. Everything else preferably should be nonstructural and easily demountable, if possible.

The space between the underside of suspended ceilings and the top of floor or roof structures must typically contain many elements besides those of the basic construction. This represents a situation requiring major coordination for the integration of the space needs for the elements of the structural, HVAC, electrical, communication, lighting, and fire-fighting systems. A major design decision that must often be made

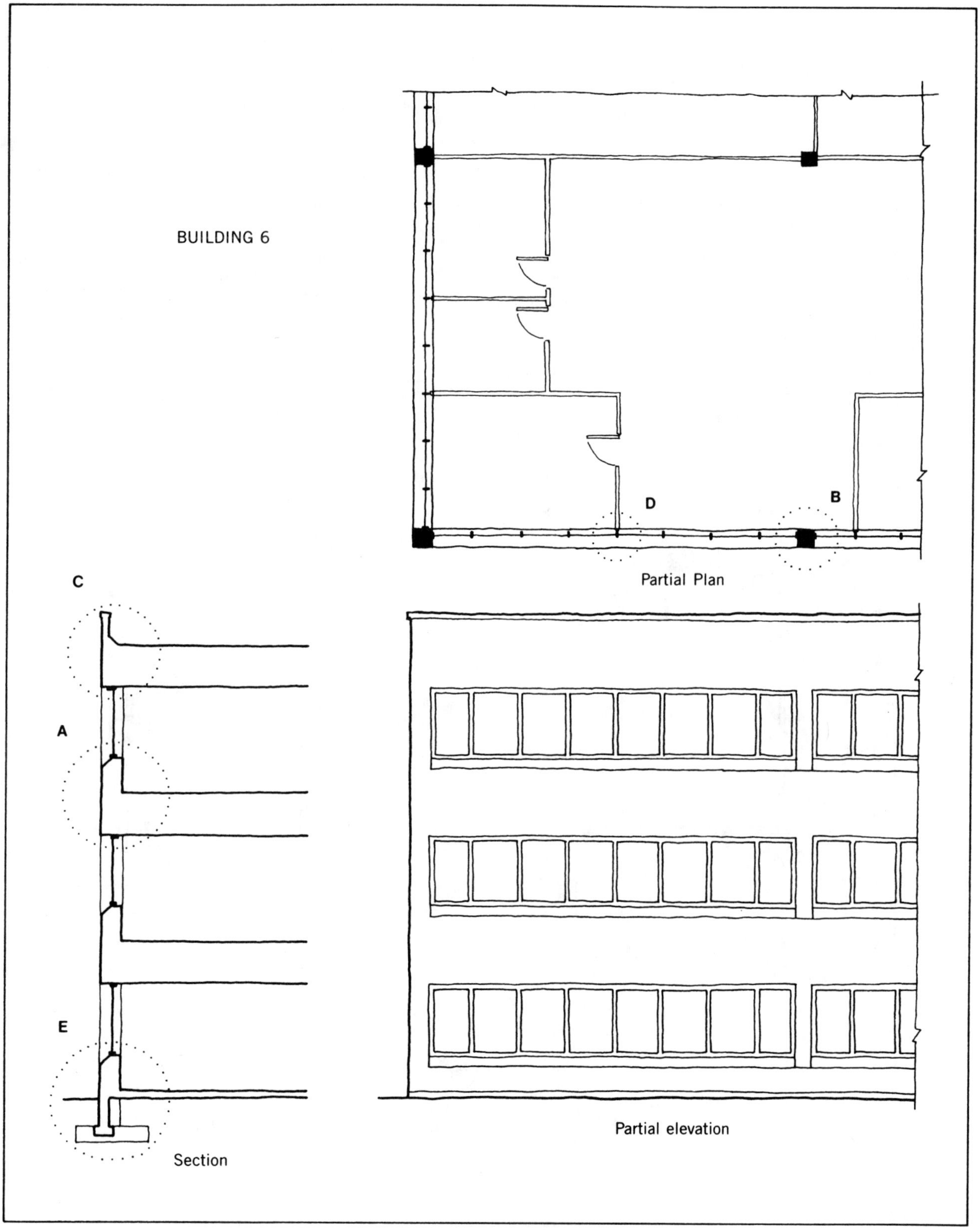

Figure 5.29 Building 6.

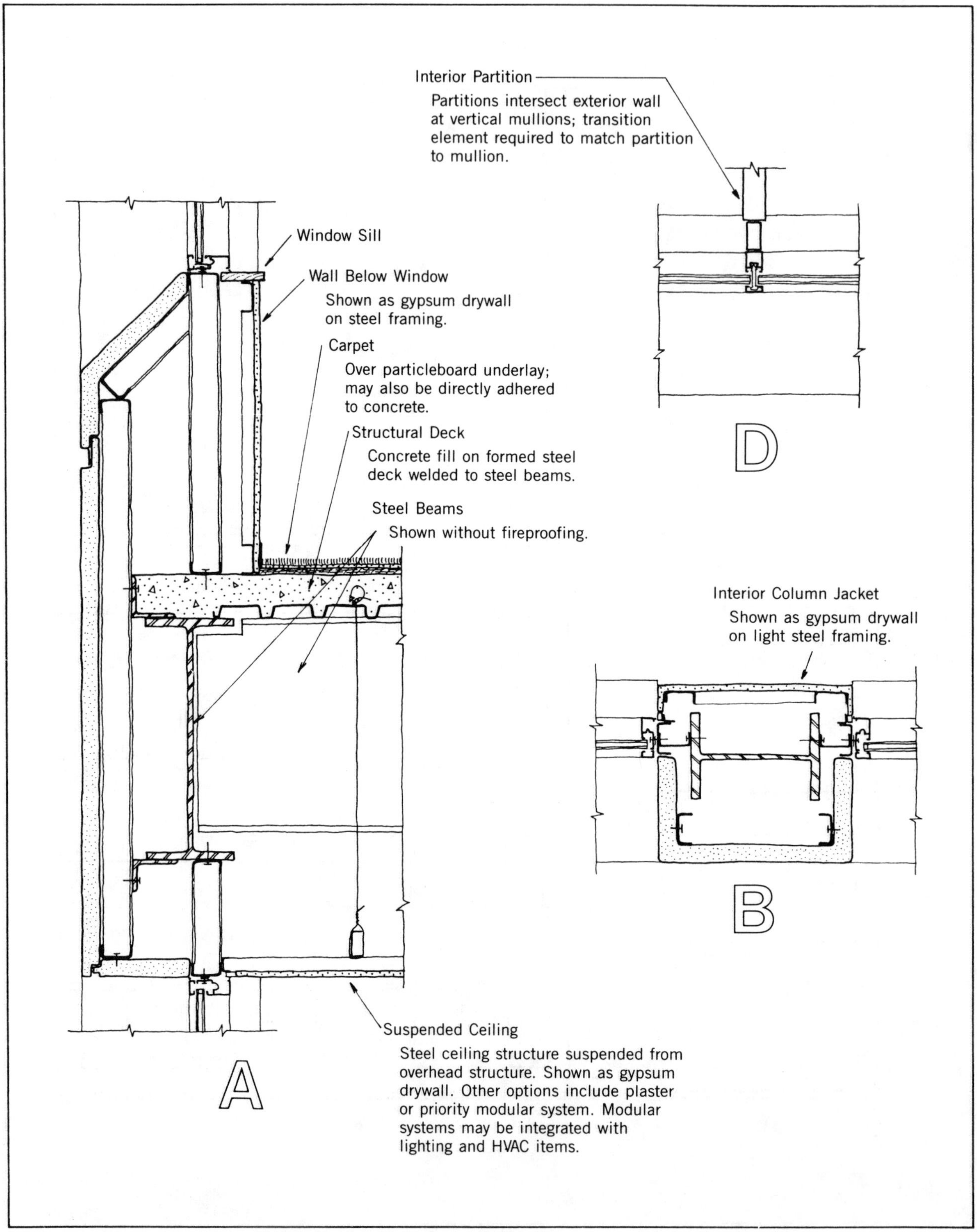

Figure 5.30 Building 6: details of the building shell and structure at the office floors.

very early in the design process is that of the overall dimension of the space required for this collection of elements. Depth permitted for the spanning structure and the general level-to-level vertical building height will be established—and not easy to change later, if the detailed design of any of the enclosed systems indicates a need for more space.

Generous provision of the space for building elements will make the work of the designers of the various subsystems easier, but the overall effects on the building design must be considered. Extra height for the exterior walls and permanent interior walls, as well as for stairs, elevators, and service risers, all result in additional cost, making tight control of the level-to-level distance very important.

Figure 5.31 shows the general scheme and some details for the HVAC system for the office areas. A ceiling-mounted air system delivers air through registers that are integrated with the general modular ceiling system. Return air to the circulating system is pulled through the ceiling, using the enclosed interstitial space between the ceiling and the structure above as a general plenum space (one big duct). Air also enters this space from grilles in the corridors. Additional details for the ceiling units are shown in Figure 5.32.

The air-handling equipment and heat-dissipating units for the air-conditioning system are located on the rooftop. Supply and return air for the offices is fed through a vertical duct shaft in the building core. The air-conditioning chiller and the furnace and hot-water boiler are in the partial basement at the building core. Hot and cold water is circulated between the basement and the equipmemt on the rooftop through pipe chases in the core.

The chilled and heated water is also fed to two other systems. One system consists of a series of *zone reconditioners* that are located on the office floors. These serve specific limited areas of the floor and perform minor modification of the temperature of the air generally circulating in the ducts, responding to the varying needs at different points on a single floor. Thus, at any given time, extra heat may be supplied on the shady side of the building at the same time that slightly cooled air is being supplied on the sunny side.

The other system fed by the dual piped water system consists of fin tube radiators along the outside walls. These are installed in the hollow space in the walls beneath the windows and feed up through grilles in the window sills, providing a wash of warm air up the inside surface of the windows. This reduces condensation and frosting on the windows during cold weather and generally adds an extra dose of heat at the cold outside walls.

This dual heating system—perimeter radiators and circulating air—is common in colder climates. The air system is generally necessary for ventilation anyway, and it is also most likely the choice for air conditioning (cooling) now provided in just about all climates.

As with other buildings, special systems may be provided for HVAC service of other building spaces. The building rest rooms will most likely have a continuously operating ventilation exhaust system, with undercut or louvered doors pulling air in from corridors to assure a one-way airflow between rest rooms and their connecting corridors.

Figure 5.33 shows some details for a general illumination system for the office areas. This system uses recessed light fixtures integrated with the general ceiling surfacing and other elements in the ceiling location. In addition to the HVAC elements (see Figure 5.32), this may include fire sprinklers, speakers for an audio system, signs, and smoke detectors. Various priority systems are available, with varying degrees of integration of these elements in a total system.

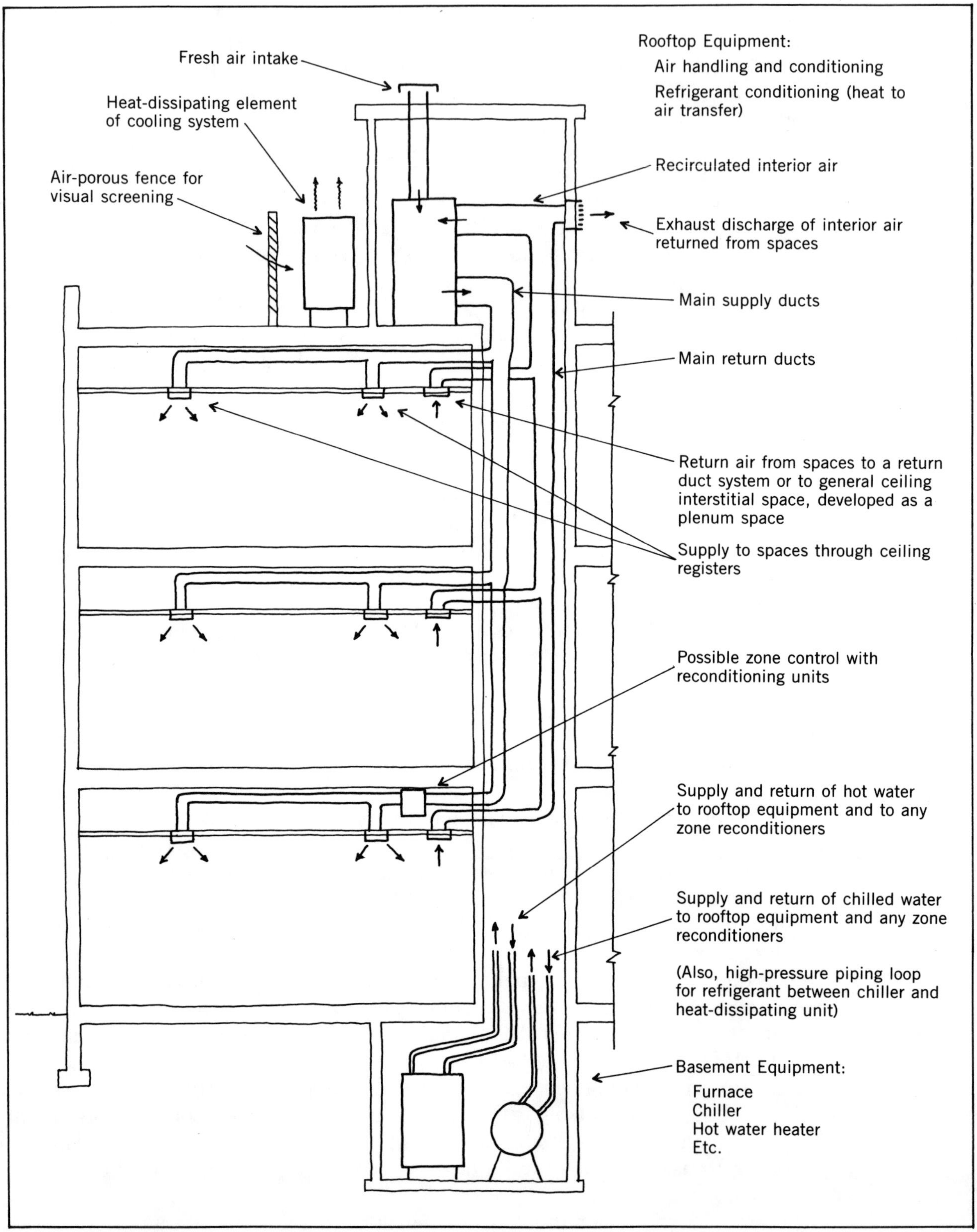

Figure 5.31 Building 6: general scheme and details for the building HVAC system.

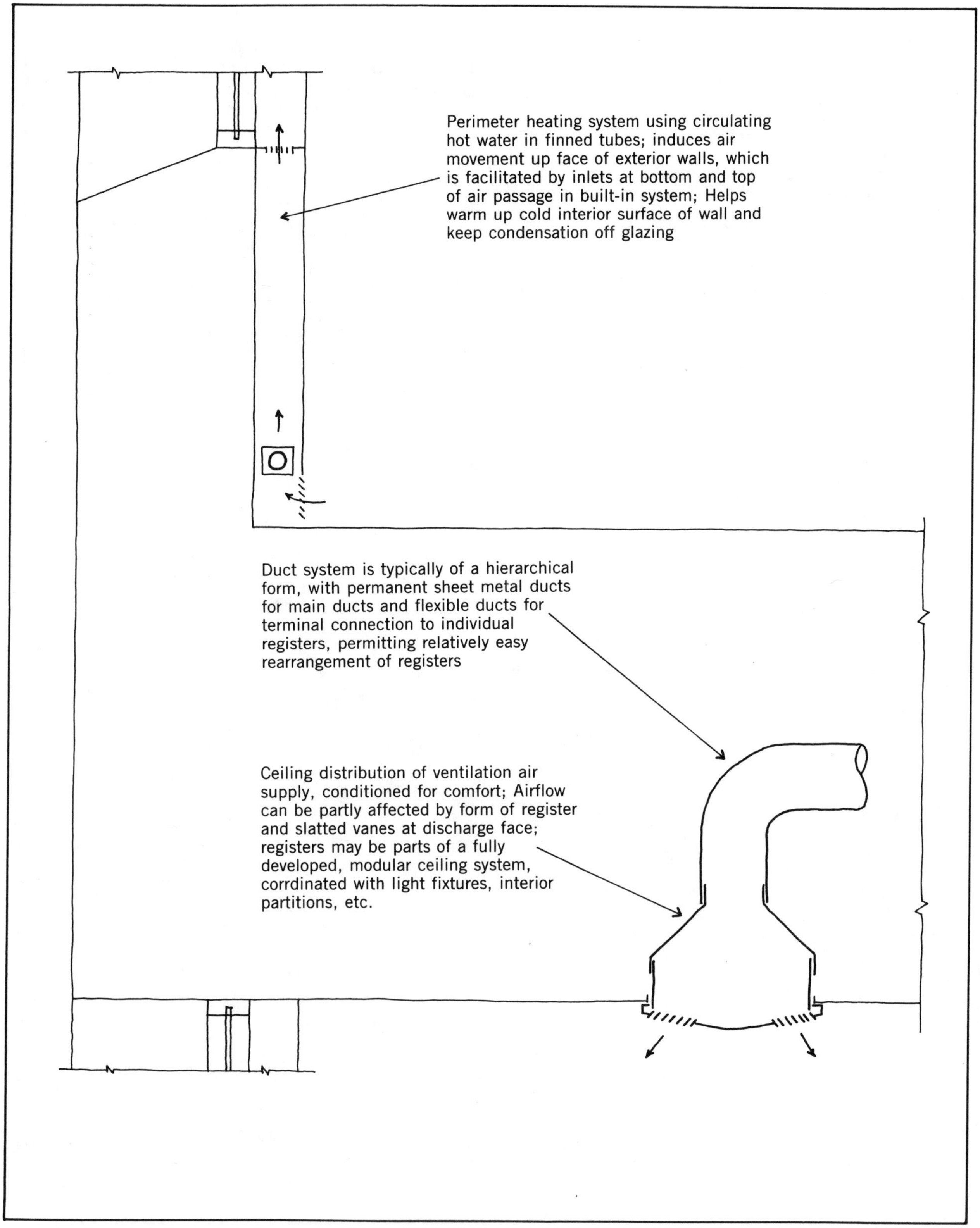

Figure 5.32 Building 6: details for the HVAC system at the office floors.

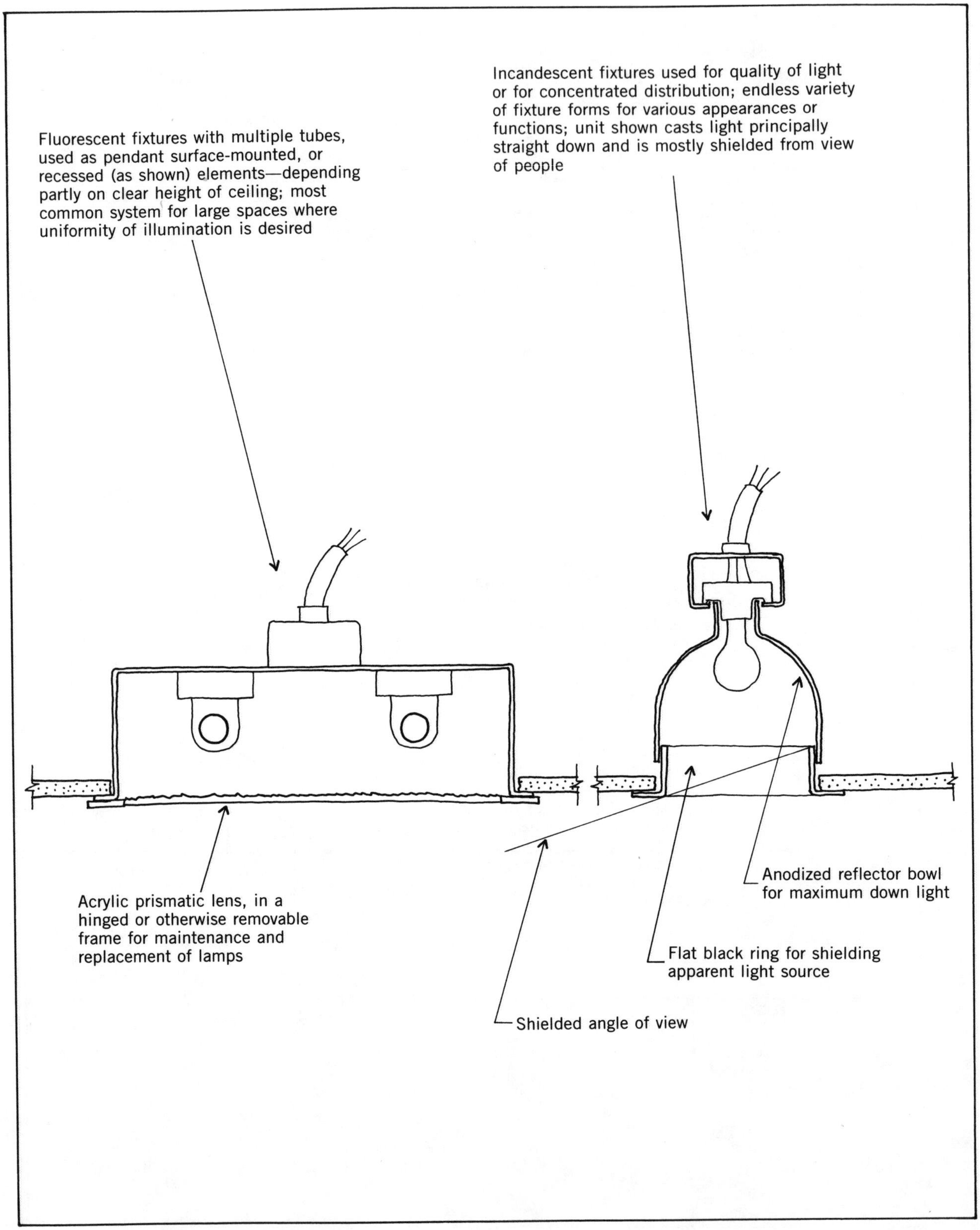

Figure 5.33 Building 6: details for the general illumination system at the office floors.

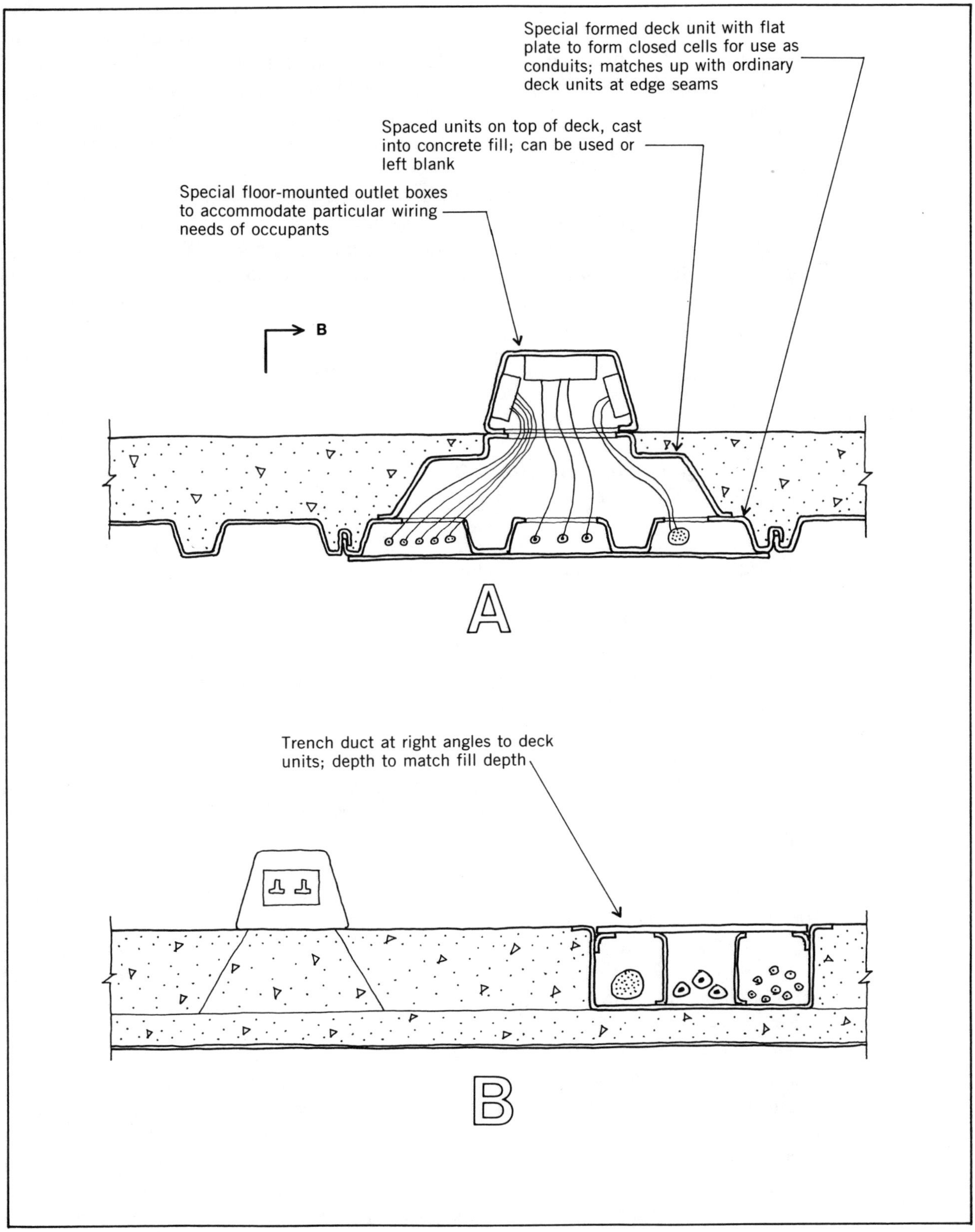

Figure 5.34 Building 6: details for the electrical power system at the office floors.

Figure 5.34 shows some details for the development of wiring distribution through a system built into the floor structure. These systems were widely used in the past to provide a form of modular, modifiable attachment for both electrical power and telephones at random locations on the floors. This random attachment is an ongoing necessity, but there are now various other options for responding to the needs. One such option is that shown for building 7, involving use of an elevated platform floor structure.

The building core is a busy spot, with risers for the various service systems, as well as the rest room, stairs, and elevtors. Tight planning and coordination of design work is required here, with many separate designers having concerns for the space use. Some details for an elevator system are shown for building 7 in Figure 5.39. A similar system would probably be used here, although service to the basement area might not be necessary.

5.7 BUILDING 7

Ten-Story Office Building
Concrete Frame, Metal Curtain Wall

This building (see Figure 5.35) is not exactly high rise, but it is enough taller than building 6 to result in some narrowing of choices for the structure and some different planning concerns. It is unlikely that anything other than a steel or concrete frame would be used for this building.

The type of curtain wall used here (see Figure 5.36) is one that forms a continuous skin outside the building structure. For comparison, see the plan and wall details for building 6, in which the columns are enclosed by the thick exterior wall at the floor level.

Formation of the curtain wall is achieved using the "stick" system, in which vertical structural mullions are developed like a stud system to provide basic support for the rest of the wall. Windows continue in horizontal rows past the columns, so the building exterior surface does not reveal the pattern of the column spacing as it does in building 6.

Spandrell covers are placed in the same plane with the windows and have glass on the exterior surface. A continuous all-glass wall is thus produced, and if dark or reflective glazing is used, the difference between glazing and spandrel panels may not be apparent in daylight. The spandrels thus also become obscured, and a generally undelineated structural surface results, with neither vertical nor horizontal structural references visually apparent.

The plans, sections, and details reveal the use of the sitecast concrete structure. The horizontal spanning system consists of a solid two-way-spanning slab, with beams used only at the edges and at large openings on the interior. One advantage of this system is the possibility for reduction of the space enclosed by the ceiling below and the structure above, as beams do not regularly project into this space as with ordinary slab and beam framing. Thus the space available for ducts and other service elements is that bounded by the top of the ceiling system below and the underside of the slab above.

Development of the suspended ceiling is shown in Figure 5.36. With this structure, it is also conceivable that the suspended ceiling might be eliminated, as the thick concrete slab does not require other fire protection. Thus the underside of the

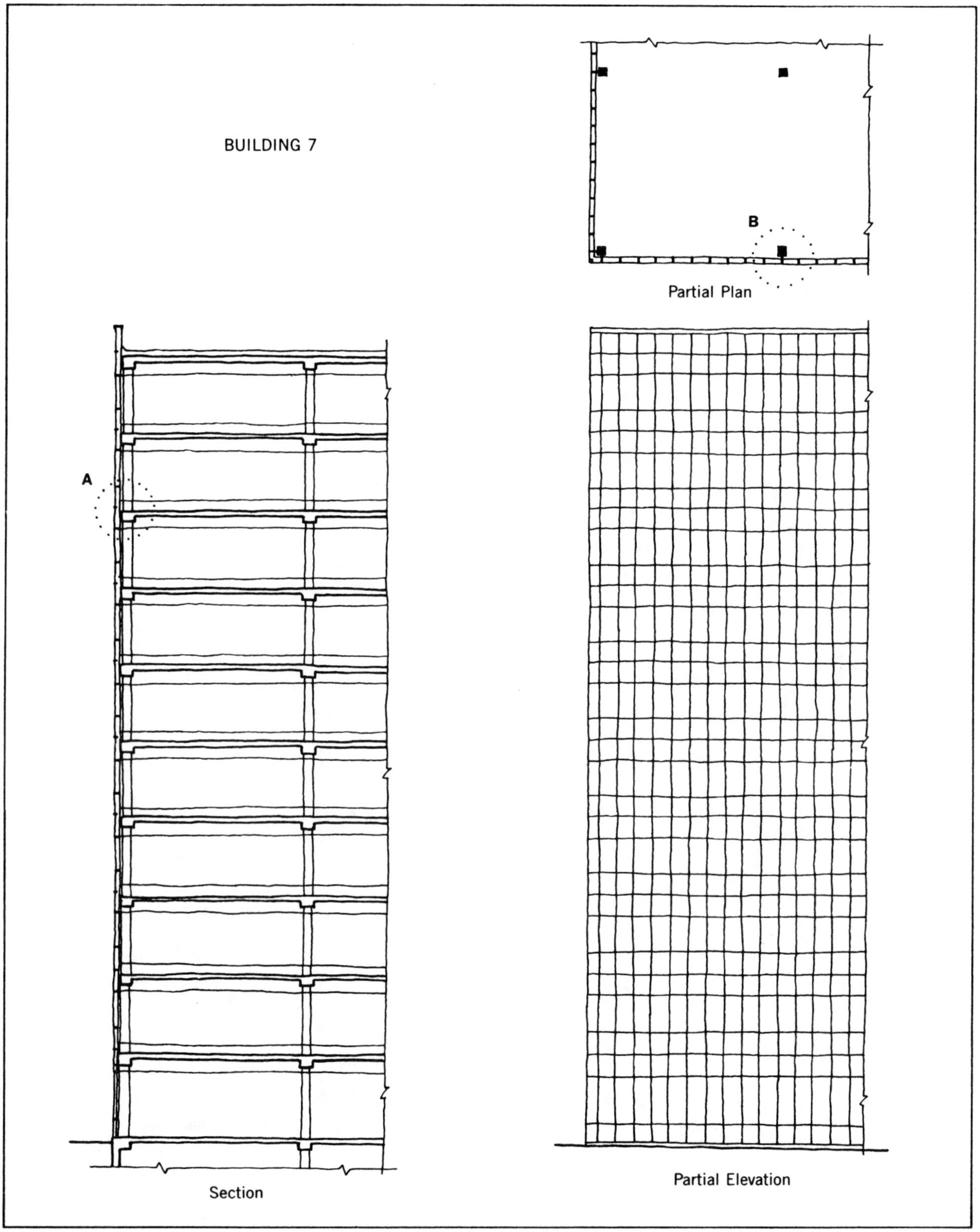

Figure 5.35 Building 7.

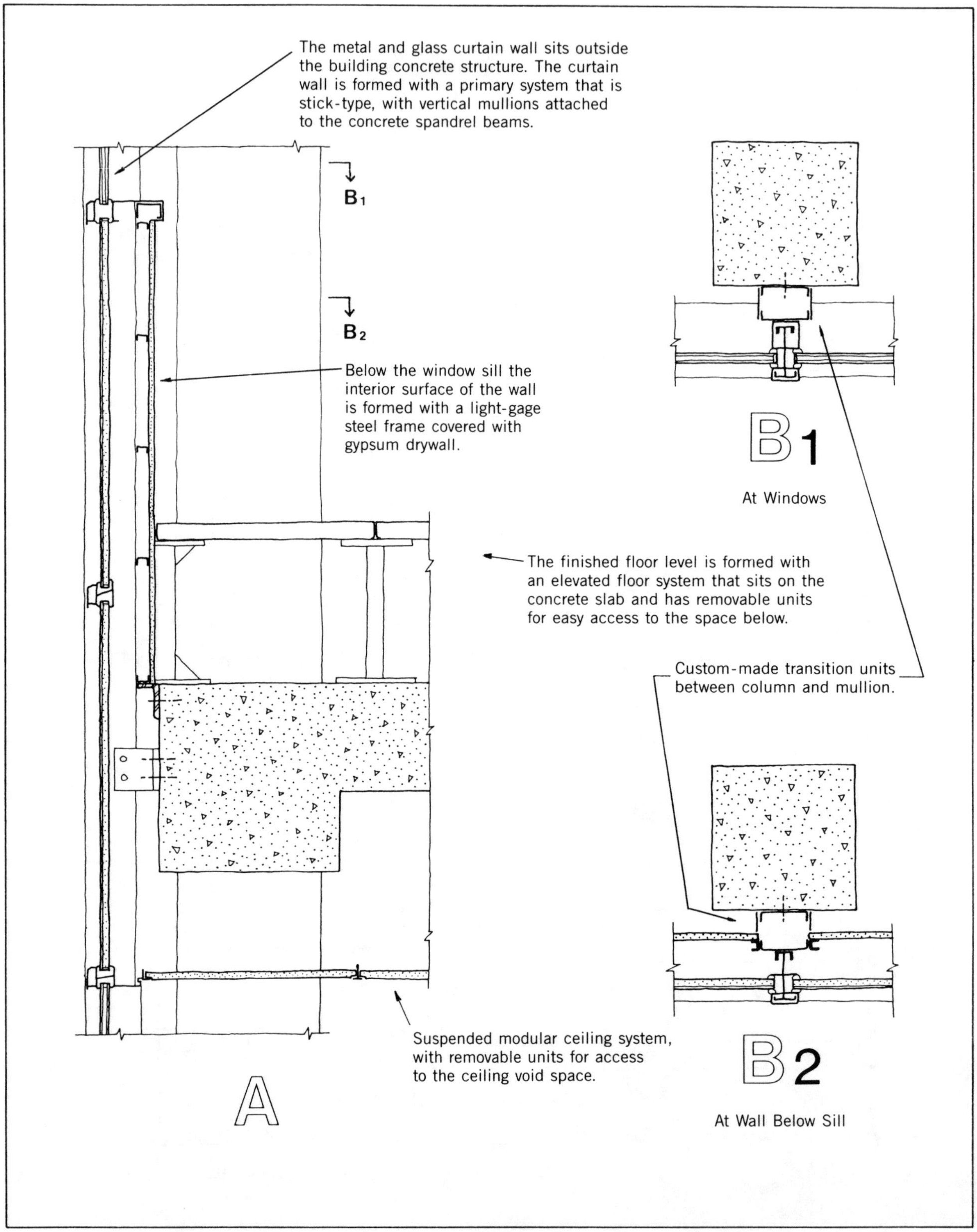

Figure 5.36 Building 7: details of the building shell and structure at the office floors.

structure could be exposed and the necessary overhead services hung out for display—a high-tech style of design popular with some designers.

For this building, the drawings also indicate the use of an elevated floor system, consisting of a platform working floor surface developed by a system of modular panels supported by closely spaced short posts. This creates an interstitial space that can be used to incorporate various building service elements—most commonly the wiring for power, phones, computer networks, and other signaling systems. For occupancies in which major use of communication and computing systems is made—and frequently modified—this is a highly useful construction.

Figure 5.37 shows some details for an HVAC system for a typical office floor. Exploiting the potential for an exposed structure here, the interior is developed without the usual suspended ceiling. While the true ceiling is the lightly treated underside of the concrete structure (probably getting just a coat of paint), an implied ceiling is visually created by the level at which most of the suspended fixtures and equipment are placed.

The exposure of the overhead service elements may be visualized as creating a freer condition for their installation. It is true, of course, that they are not strictly bound by the hard definition of a structured interstitial space and do not need to dodge and fit into a constructed ceiling system. However, as much as their exposure may create a sense of reality, it is quite likely that much greater care will be required for their installation. This is not a basement or a warehouse, where there is virtually no concern for appearance. This is a *high-tech* interior design style, with a very controlled appearance, akin to a person wearing intentionally trashed designer jeans—not bargain store jeans, mind you!

With both the possibilities for interstitial space (the suspended ceiling and the elevated floor) at their disposal, the interior designers and the engineers for the various service systems have some options for location of items. Lighting for general illumination and fire sprinklers must be placed at the usual ceiling locations. However, wiring, ducts, and various items may be located in either the ceiling or underfloor positions. This flexibility can be exploited for real practical purposes, or it may be played with for freedom of expression in the interior design.

Figure 5.38 shows some details for the elevated floor system, using the priority elements of one manufacturer. The basic manufactured system provides for the general floor structure and some special conditions, but some expedient detailing or matching construction is usually needed for the full development of the building floor system. In this case, in order to keep the whole floor at one level, a special system is used for the corridors. Similar solutions may be developed for the floors of rest rooms, stair landings, elevator lobbies, and other locations on a typical office floor. There are many options in this situation, and the suppliers of a particular priority system may well have their own solutions for most common occurrences in order to further promote use of their products.

Figure 5.39 shows some details for the elevator system, which uses a high-speed, overhead suspended operating system. To serve the lowest levels of the basement parking, the construction must be extended considerably below the level of the floor of the lower basement, requiring some special considerations in development of the building foundations in the immediate vicinity. For the typical installation, there is also the need to extend the construction almost two stories above the highest level served by the elevators. The latter situation normally produces some form of "penthouse" structure on the building rooftop, unless some considerable attic space is created by the development of the building roof.

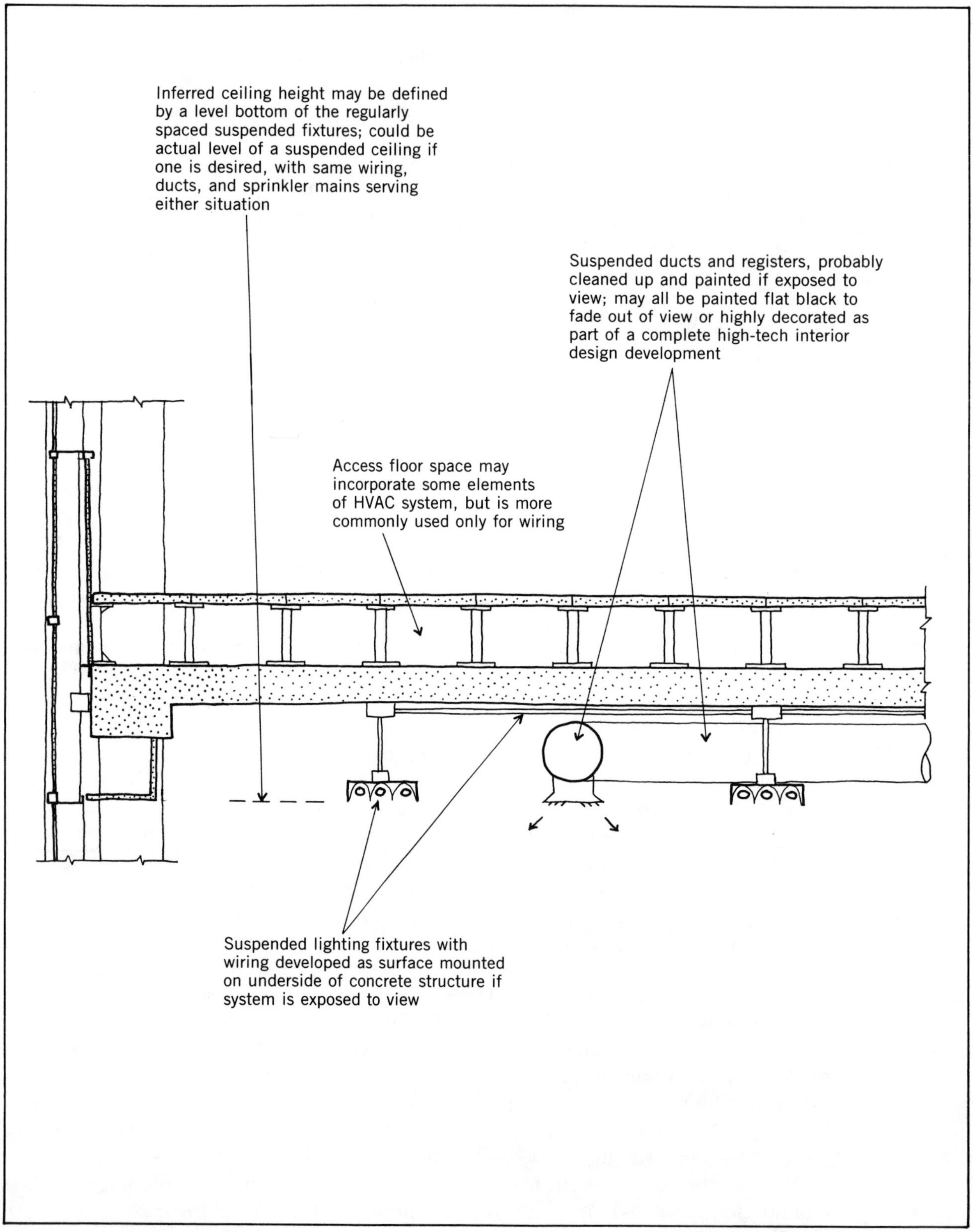

Figure 5.37 Building 7: General scheme and details for the HVAC systems.

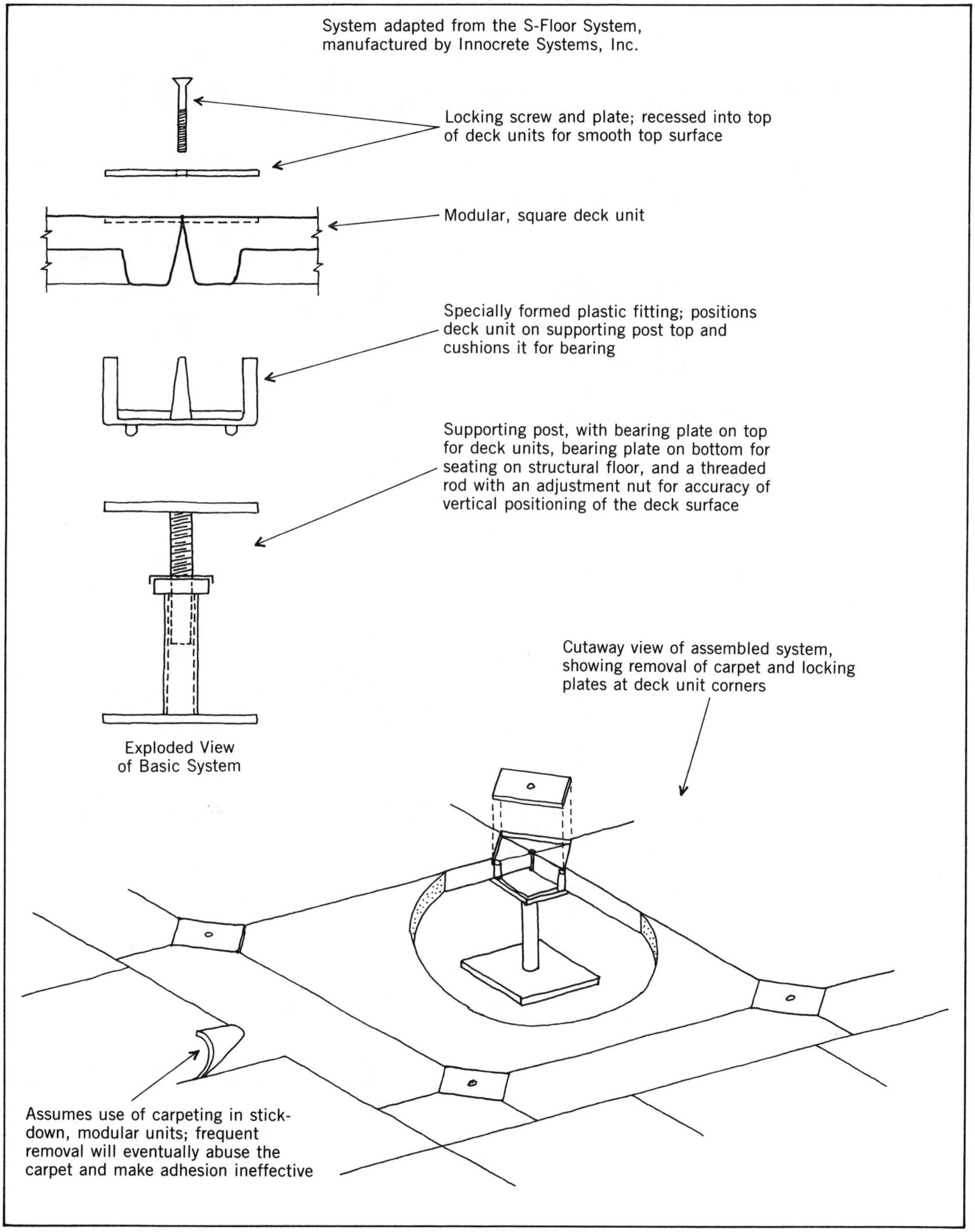

Figure 5.38 Building 7: details for the elevated floor system.

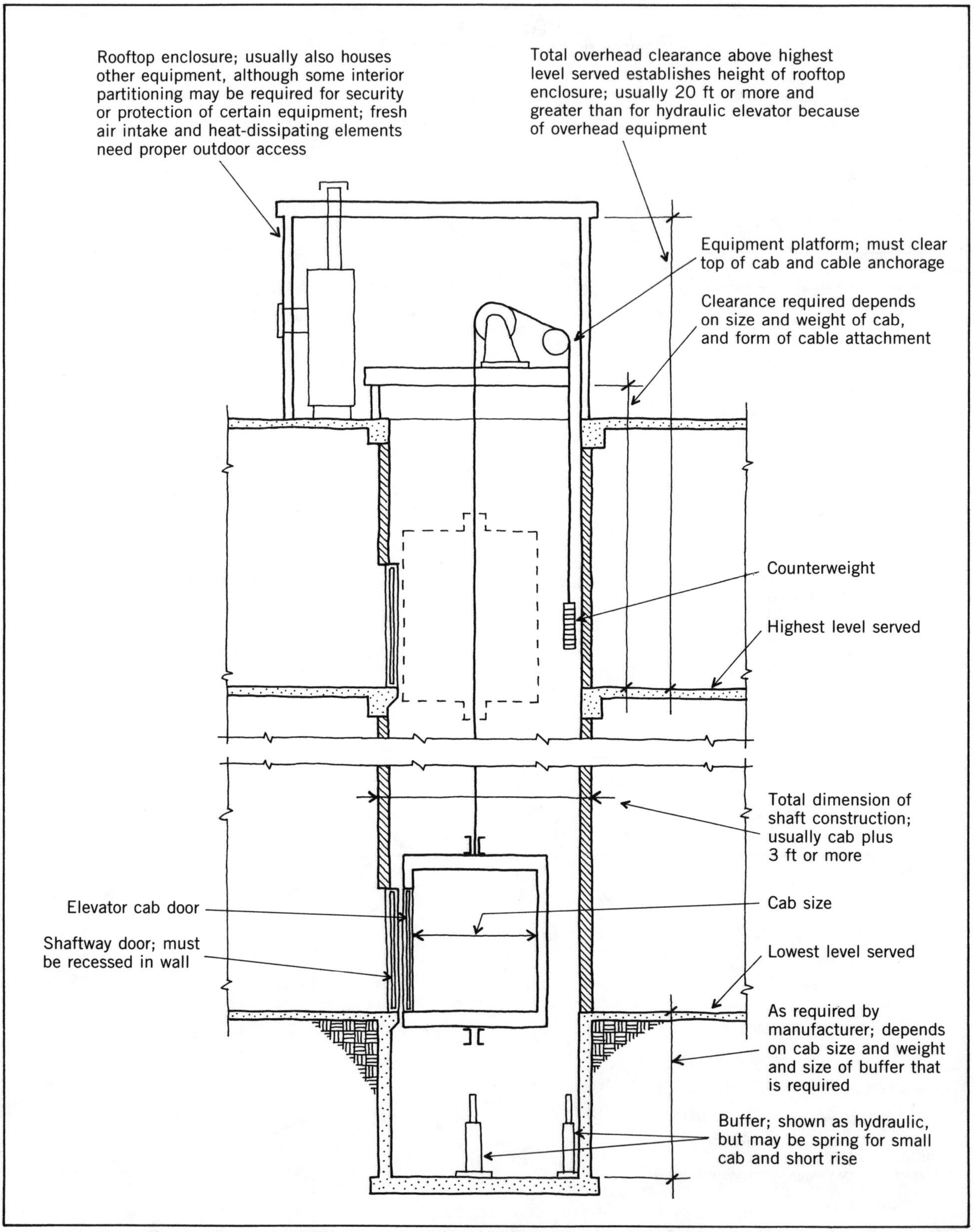

Figure 5.39 Building 7: details for the elevator system.

When the elevators are placed in a "core" group with other permanent service elements (notably the vertical ducts for the HVAC system), it is common to use a general rooftop structure to enclose some of the necessary rooftop equipment for other services. This is a chronic problem for the flat-roofed building, and design responses vary considerably.

The solutions shown here for buildings 6 and 7 are not particularly unique, and they show some range of possibilities for the typical multistory building. There are, however, many other possibilities that may well be more appropriate for other occupancies or for particular climate conditions, budgetary limitations, local building code restrictions, and so on.

The building shown here is a quite common occurrence in tight urban locations, where site limitations and local zoning requirements conspire to require some underground parking. A service system problem of some consequence in this situation is that of the provision of a ventilating system for the parking facility (see Figure 5.40). The basic concern is for removal of gas fumes, which lay at floor level.

While local codes may vary, the usual requirements for an exhaust system for enclosed parking garages include the following provisions.

Positioning of intake grilles for the ducted exhaust system at not more than 18 in. above the floor, at spacing of not more than some distance (50 ft or less) along the entire perimeter of enclosed spaces and near corners or other spots where gas may be trapped

Total air change per hour at a rate considerably greater than that for typical occupied spaces

Air change that is 100% fresh air, with exhaust dumped out at points not likely to permit the toxic funes to disturb people on or off the site.

The typical result is a system with some enormous fans (with enormous noise and vibration) which must usually operate continuously, with large main ducts and good-sized intake and exhaust grilles. This results in the strong preference for an open parking structure (no actual enclosure for an HVAC system to have to deal with) whenever possible. If planning of the building and site simply does not allow this, the exhaust system must be paid for and incorporated into the construction and space planning for the building and site.

5.8 BUILDING 8

Sports Arena, Longspan Roof
Steel Two-Way Truss System

This is a medium-size sports arena, possibly big enough for a swim stadium or a basketball court (see Figure 5.41). Options for the structure here are strongly related to the desired building form. Functional planning requirements derive from the specific activities to be housed and from the seating, internal traffic, overhead clearance, and exit and entrance arrangements.

In spite of all of the requirements, there is usually some room for consideration of a range of alternatives for the general building plan and overall building form. Choice of the system shown here relates to a commitment to a square plan and a flat roof profile. Other choices for the structure may permit more flexibility in the building

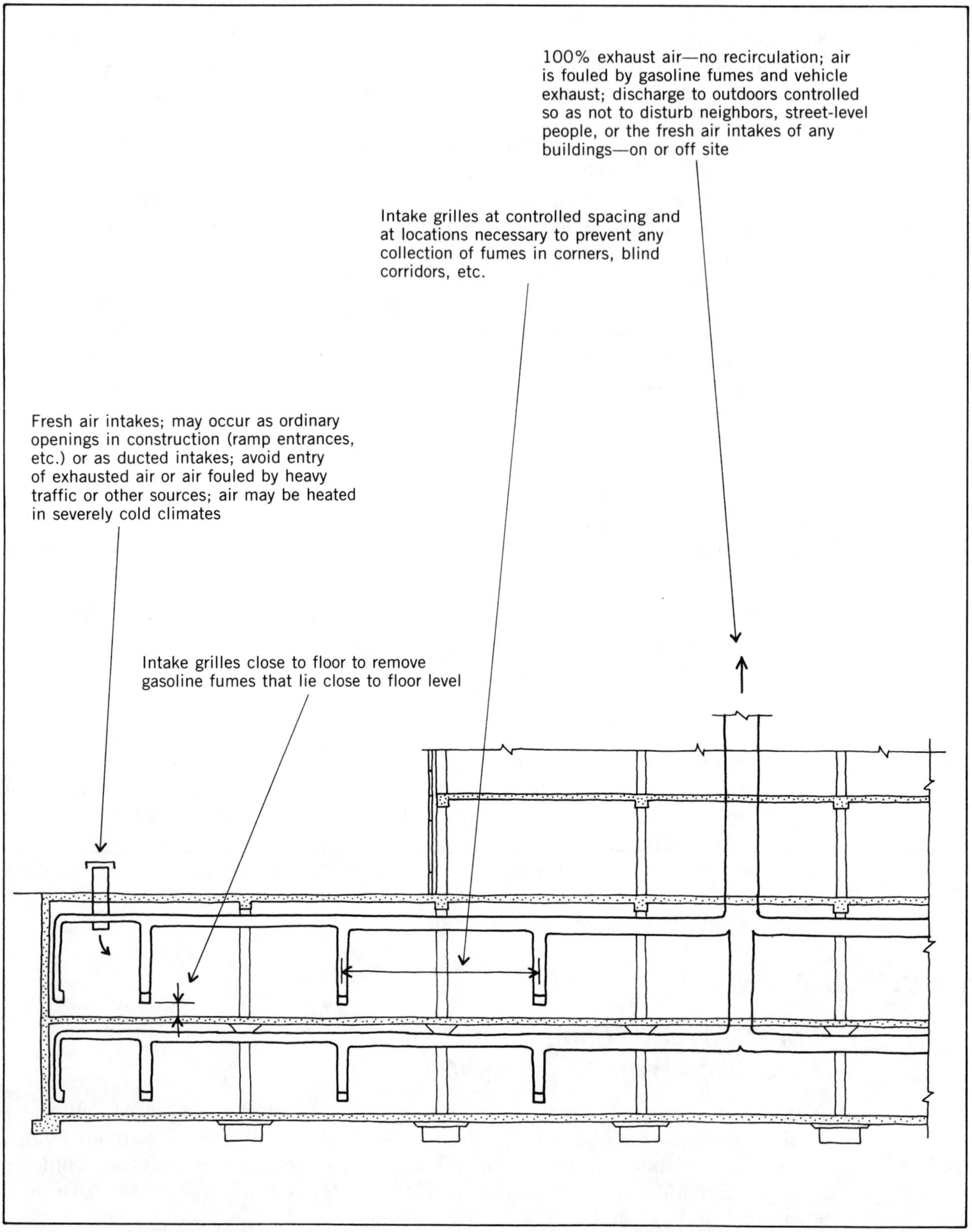

Figure 5.40 Building 7: details for the ventilation system for the parking garage.

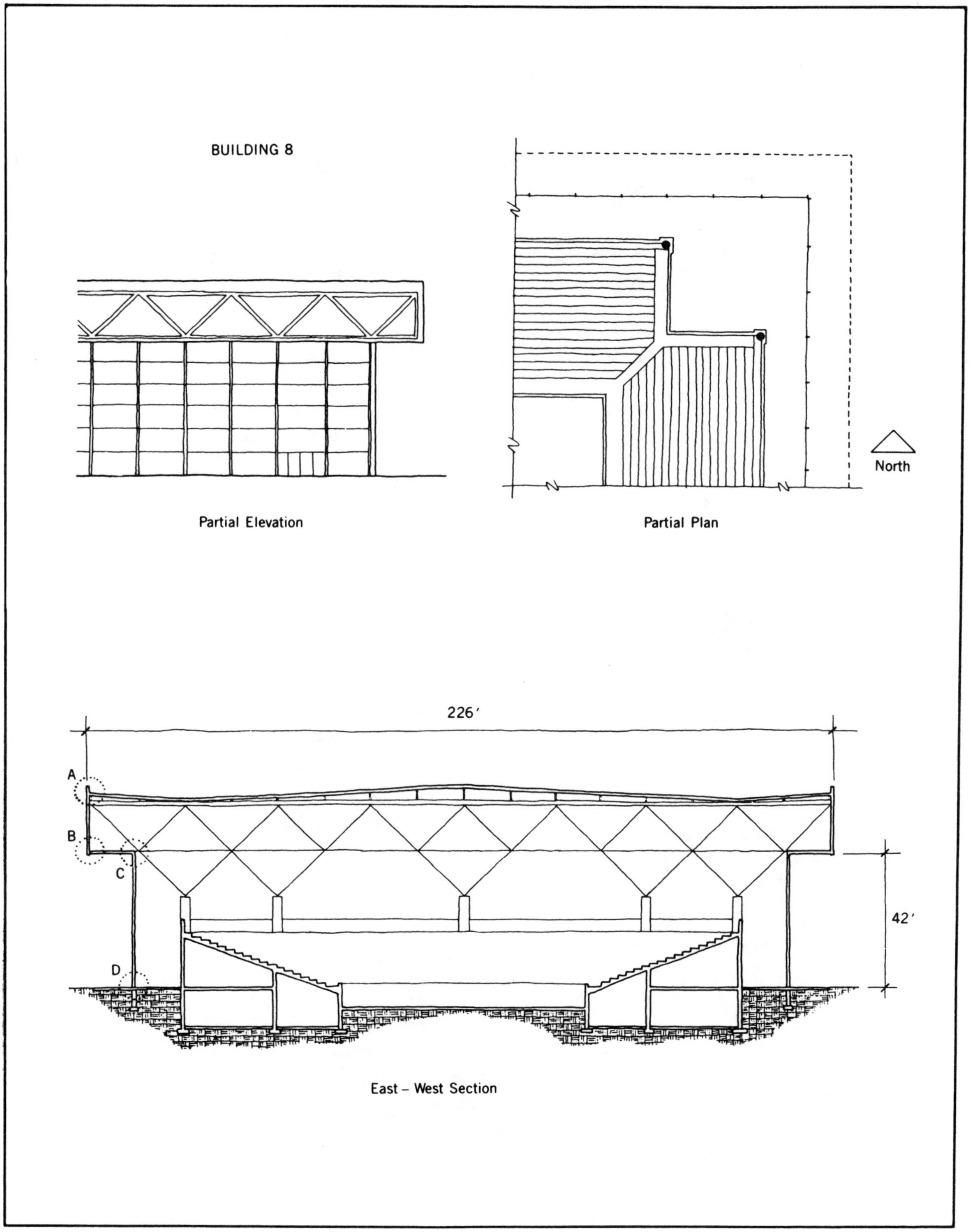

Figure 5.41 Building 8.

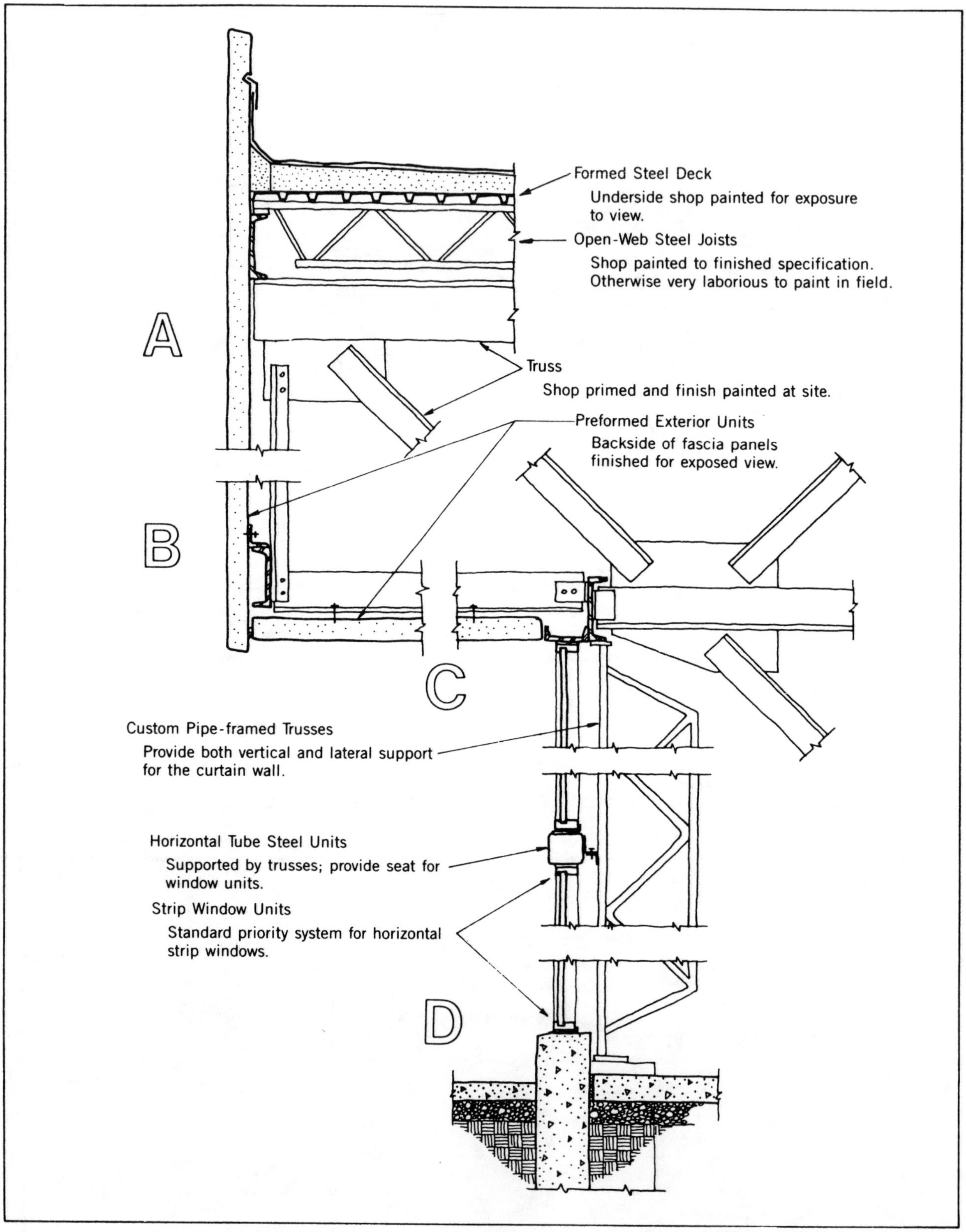

Figure 5.42 Building 8: details of the building shell.

form or also limit it. Selection of a dome, for example, would require a round plan, or at least one that could be developed under the hemispherical enclosure. Need for an oblong plan would limit the possibility for a two-way spanning structure.

The structure shown here uses a two-way spanning steel truss system (often called a "space frame," although the name is quite ambiguous). Basic planning with this type of structure requires the use of a module that relates to the nodal points (joints) of the truss system. Locations of supports must relate to the nodal points and any major concentrated loads should be applied at nodal points. The use of the nodal point module may be extended to general planning of the building, as has been done in this example.

There is nothing particularly unique about the construction shown here, but it cannot really be called common or standard. Even in unique buildings, however, it is common to use as many standard products as possible. Thus the roof structure on top of the truss and the curtain wall system for the exterior walls use off-the-shelf products (see Figure 5.42).

It is possible that the truss system shown here could be produced with various priority systems, and it is generally advisable to pursue these when beginning the design of such a structure. The details shown here indicate an ordinary truss system using rolled steel shapes and steel plate gussets, a possibility among various options. For the exposed structure another popular form is that produced with steel pipes—in the manner of a giant bicycle frame.

In addition to the longspan structure in this case, there is also the major problem in developing the 42-ft high curtain wall. Braced laterally at only the top and bottom, this is a 42-ft span structure sustaining wind pressure as a major load. With even modest wind conditions, producing pressures in the 20-psf range, this requires some major vertical structure to span the 42 ft.

In this example, the fascia of the roof trusses, the soffit of the overhang, and the curtain wall surface are all developed with products ordinarily available for curtain wall construction. The vertical span of the tall wall is developed by a series of custom-designed trusses that establish a major spacing module for the vertical mullions, analogous to the columns in building 7. A horizontal structure is developed to span the distance of 14 ft between the vertical trusses—steel structural tubing in this case. The steel tubes then provide both vertical support and horizontal bracing for the 3.5-ft-wide by 7-ft-high window units.

The inside of the roof and curtain wall shell and the exposed steel truss system strongly define the building interior. The truss members define a pattern that is orderly and pervading. There will, however, most likely be additional items overhead, within and possibly beneath the trusses. These may include:

Drains and piping for the roof drainage system

Ducts and registers for the building HVAC system

Wiring and fixtures for a general lighting system

Elements required for a general audio system

Signs, scoreboards, etc.

Catwalks for access to the various overhead equipment

To preserve some design order, as much of the overhead equipment as is possible should be developed within the truss modular system.

The building shell, of course, exists only to house the building activities. In this case, an essentially independent construction is developed under the large overhead enclosure to form the seating and other required facilities. The support columns for the truss are incorporated into this construction, but the separation is otherwise complete and the planning could be totally independent.

Although the size, shape, general construction, and usage are all different, buildings 5, 8, and 9 are similar in that they are generally developed as single-space enclosures, with the interior construction largely free of the enclosure. The enclosure must generally accommodate the interior activities, but what happens on the inside is subject to considerable variation. Buildings 1 and 4 are quite different in this regard, while buildings 2, 3, 6, and 7 are transitional between the two situations, accommodating some limited interior variation.

A major service problem for this building is the handling of the large building volume as a single space for ventilation and thermal control. The vertical interior height of the space makes it quite difficult to control air temperatures at specific vertical positions within the space. When the same air system is used for both heating and cooling, this is further complicated, since the natural gravity effects make the warm input air rise and the cool input air drop.

The solutions for the HVAC systems for this building must respond to a number of variables, including the following.

Climate: The form of the HVAC system may well depend on whether the predominant problem is winter heating or summer cooling. Both may be required, but one or the other is often the major concern, and the fundamental choices for equipment, duct and register form and locations, and other factors may favor one function over the other.

Building usage: This includes concerns for the range of activities to be facilitated, the predominant times of day or seasons of the year of highest use, the frequency of use (basically for continuous and daily or only for infrequent major events).

Favored users: For athletic events with spectators it must often be decided whether to favor the athletes or the spectators. The spectators have a more-or-less constant condition for comfort, based on long periods of recline with little activity. The athletes, however, may range from potentially overheated (basketball players) to potentially chilled (swimmers). It is hard to accommodate all of this.

Adding to the problem here is the need to deal with what amounts to a second building within the larger building, consisting of the enclosed spaces beneath the seating. This space has its own unique problems and particular needs for ventilation and general thermal control (locker rooms, large showers, smelly storage, etc.). This is likely to result in an essentially independent system, with connections bypassing the upper space through tunnels for air intake and exhaust—virtually like a separate, underground building. However, space for equipment and the fresh-air intakes may possibly be shared.

Figure 5.43 shows some details for a system for ventilation of the large arena space, using four separate supply air towers in the building corners, feeding a perimeter duct and overhead register system. Return air is handled through grilles in the seating structure. This system might be coupled with a low-pressure system using small registers within the seating to provide better control at the seating locations. Equipment for the system is in the concrete structure, with air intake and exhaust through tunnels connecting to terminals outside the building.

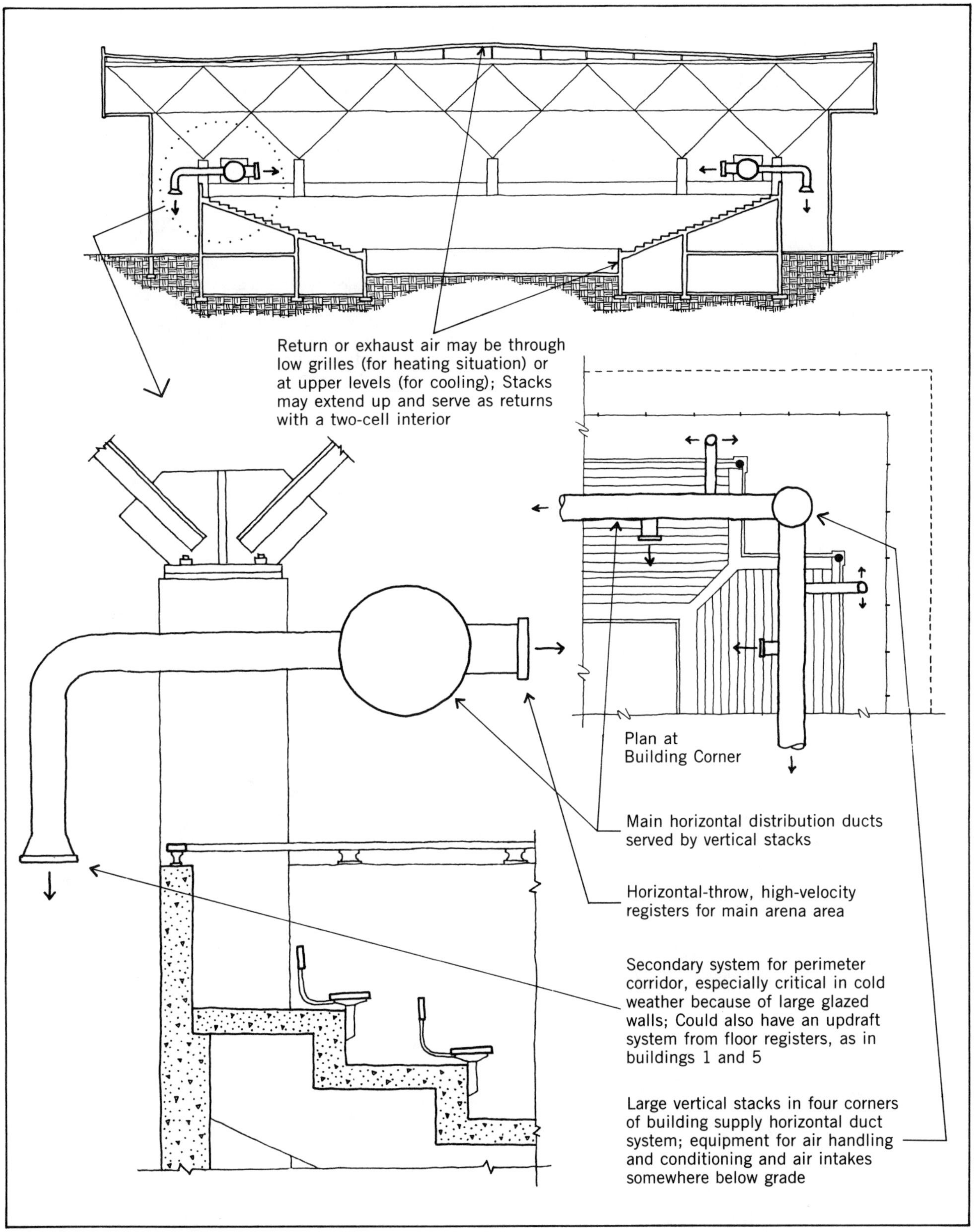

Figure 5.43 Building 8: general scheme and details for the HVAC system.

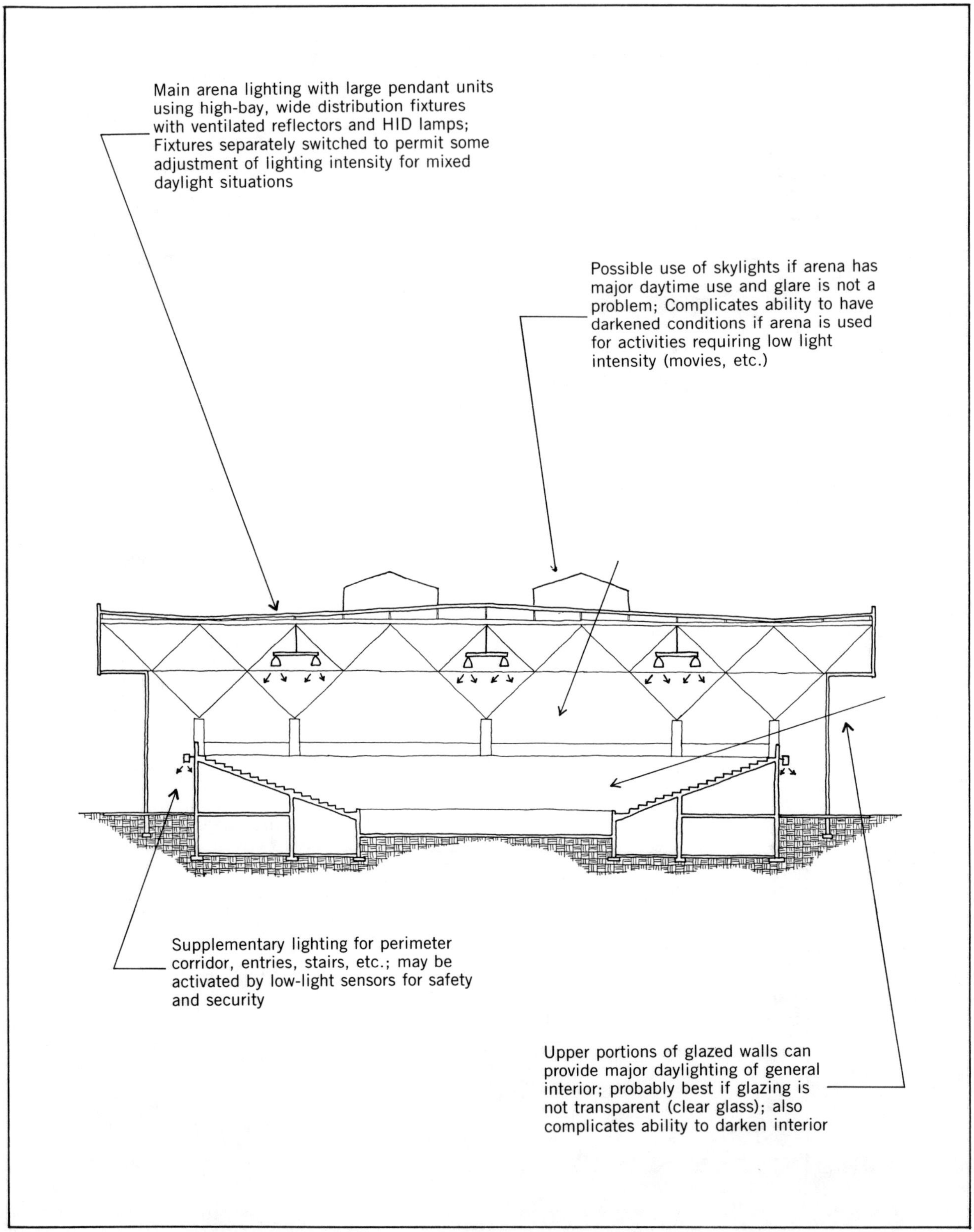

Figure 5.44 Building 8: details for the electrical lighting systems.

Another major problem for this building is the provision of a general illumination system. If usage is primarily daytime, it is likely that use would be made of natural light as much as possible. This would include input from the large glazed walls and conceivably from some skylights. However, the facility is sure to be used at night as well, and the basic system should be one that permits both conditions.

Figure 5.44 shows some details for a lighting system that uses suspended fixtures with multiple lamps. Individual lamps, as well as individual fixtures, can be separately switched to permit a considerable range for the system. During daytime, for example, the fixtures near the perimeter would probably be switched completely off.

Considerable other lighting would also be necessary, including that for the enclosed spaces beneath the seating. Special lighting for events is also likely, and a catwalk system may well be permanently installed at the level of the bottom of the trusses to facilitate this as well as maintenance of other elements within the truss system.

5.9 BUILDING 9

Open Pavilion
Wood Pole Frame and Timber Truss

This is an open-air facility of medium span, consisting only of a large roof over a small amphitheater that is built into the ground (see Figure 5.45). Foundations and vertical supports for the roof structure are provided by wood poles with their bottom ends buried in the ground. The roof structure consists of a system of timber trusses supporting wood purlins and a wood plank roof deck. The roof surface is developed with a sheet metal roofing system with standing ribs.

This is a somewhat primitive structure, with some level of roughness in the finished quality—especially the timber poles, which may be simple tree trunks with only approximately round sections, tapered profiles with changing diameters, and many knots, splits, and so on. The truss members and purlins may also be rough, but could be standard structural lumber.

The roof deck is visible only on the underside, and thus only this surface is of concern for appearance. The plank units may be solid sawn, but are increasingly likely now to be laminated, with the viewed surface chosen for appearance. General details of the wood structure are shown in Figure 5.46.

There is some analogy here with the structure for building 8, although the different materials, smaller size, and lack of an enclosing wall create considerable differences. What *is* common is the development of the interior with the exposed underside of the roof and the exposed structure. Also, as with building 8, the activity being housed is quite independent from the roof structure in terms of construction.

The only interior construction in this case is that provided for the seating, which defines the form of the amphitheater. As with building 8, this is essentially an independent structure, fitting within the overhead structure but quite distinct from it.

This building is not likely to have many services in the usual sense. There may indeed be none—not even lighting, if the usage is only during the daytime. However, some lighting is probably likely, with the form and details depending on various considerations. Figure 5.47 shows some details for a fast, cheap solution using fixtures mounted on the overhead structure with wiring in surface-mounted con-

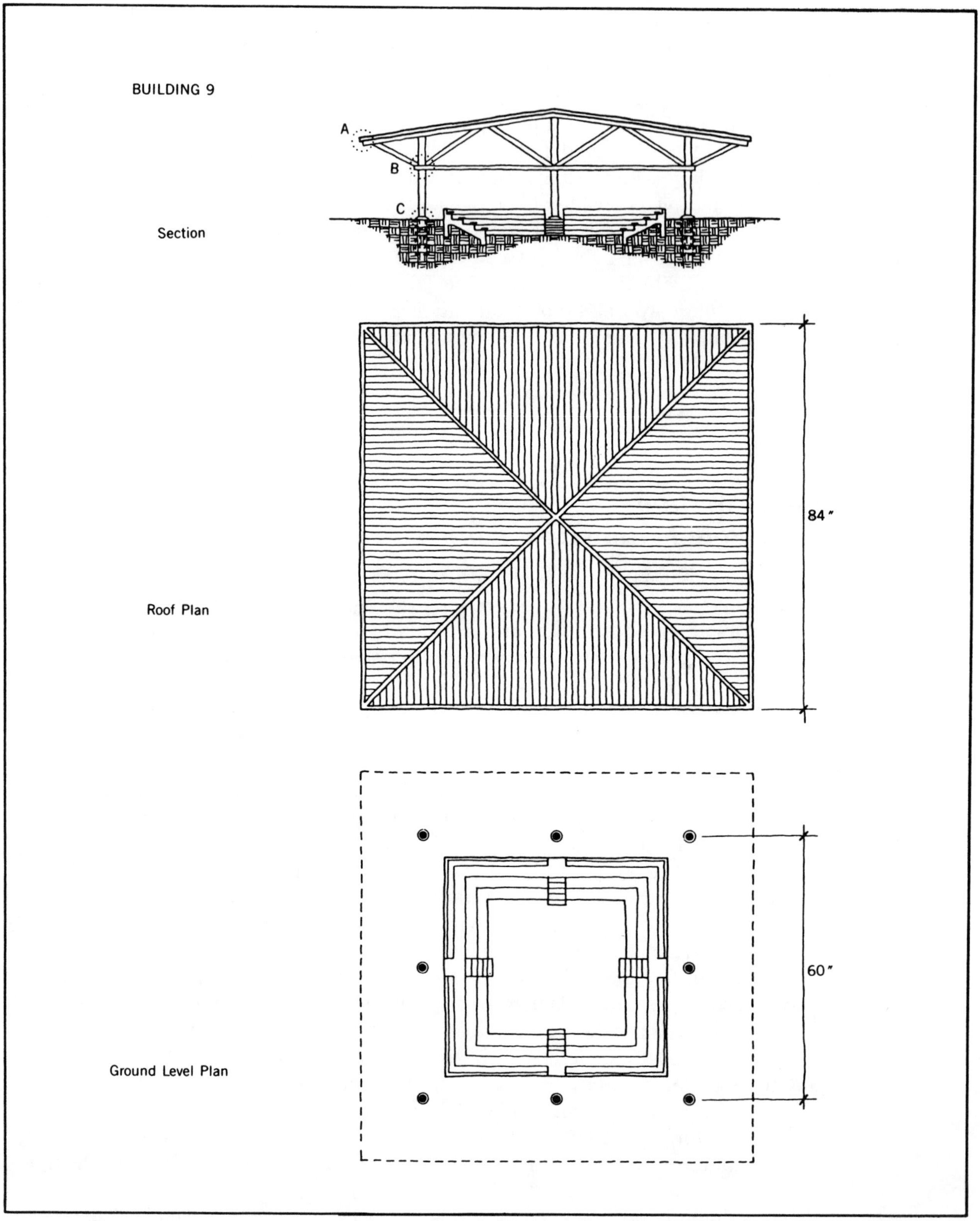

Figure 5.45 Building 9.

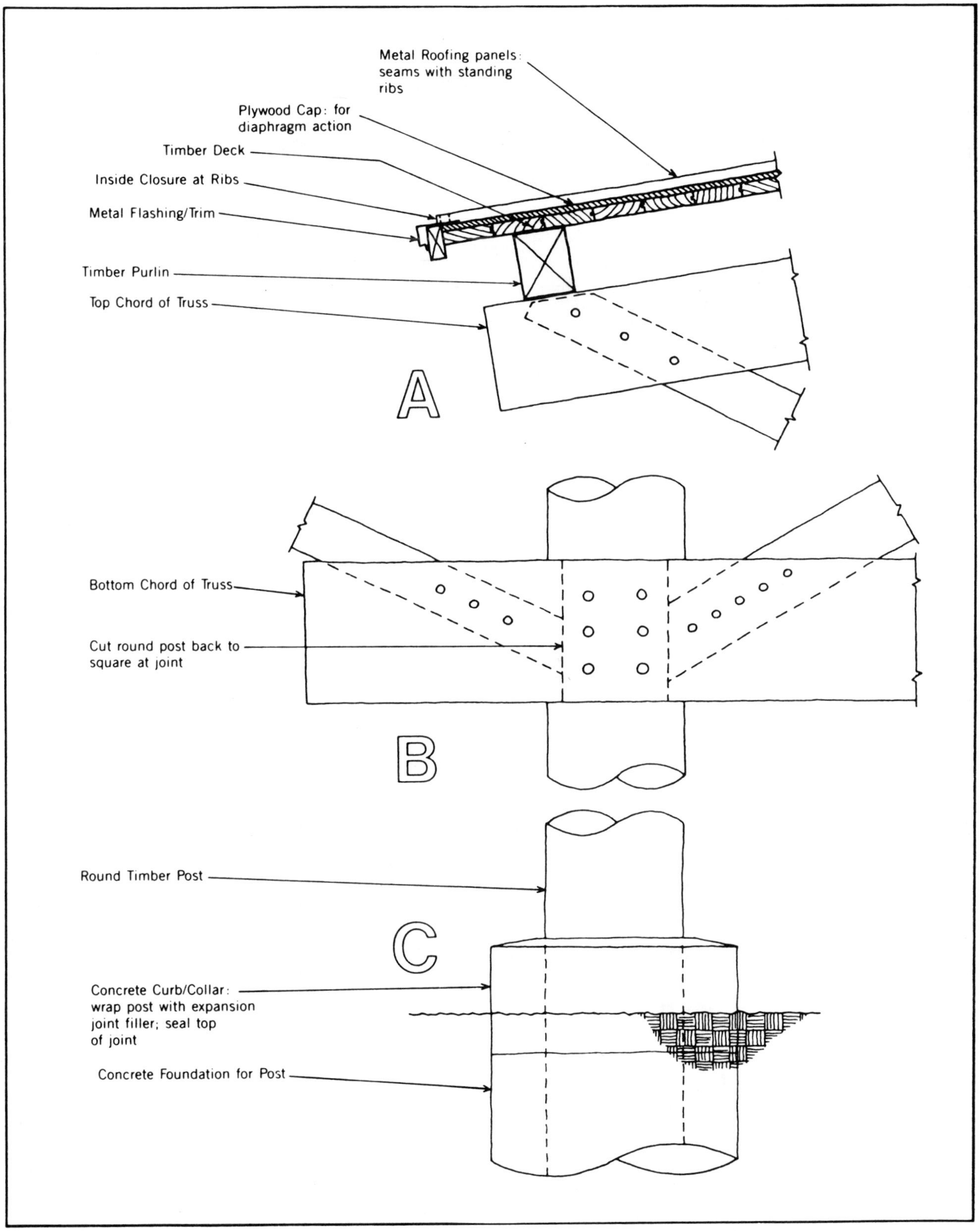

Figure 5.46 Building 9: details of the column and roof structure.

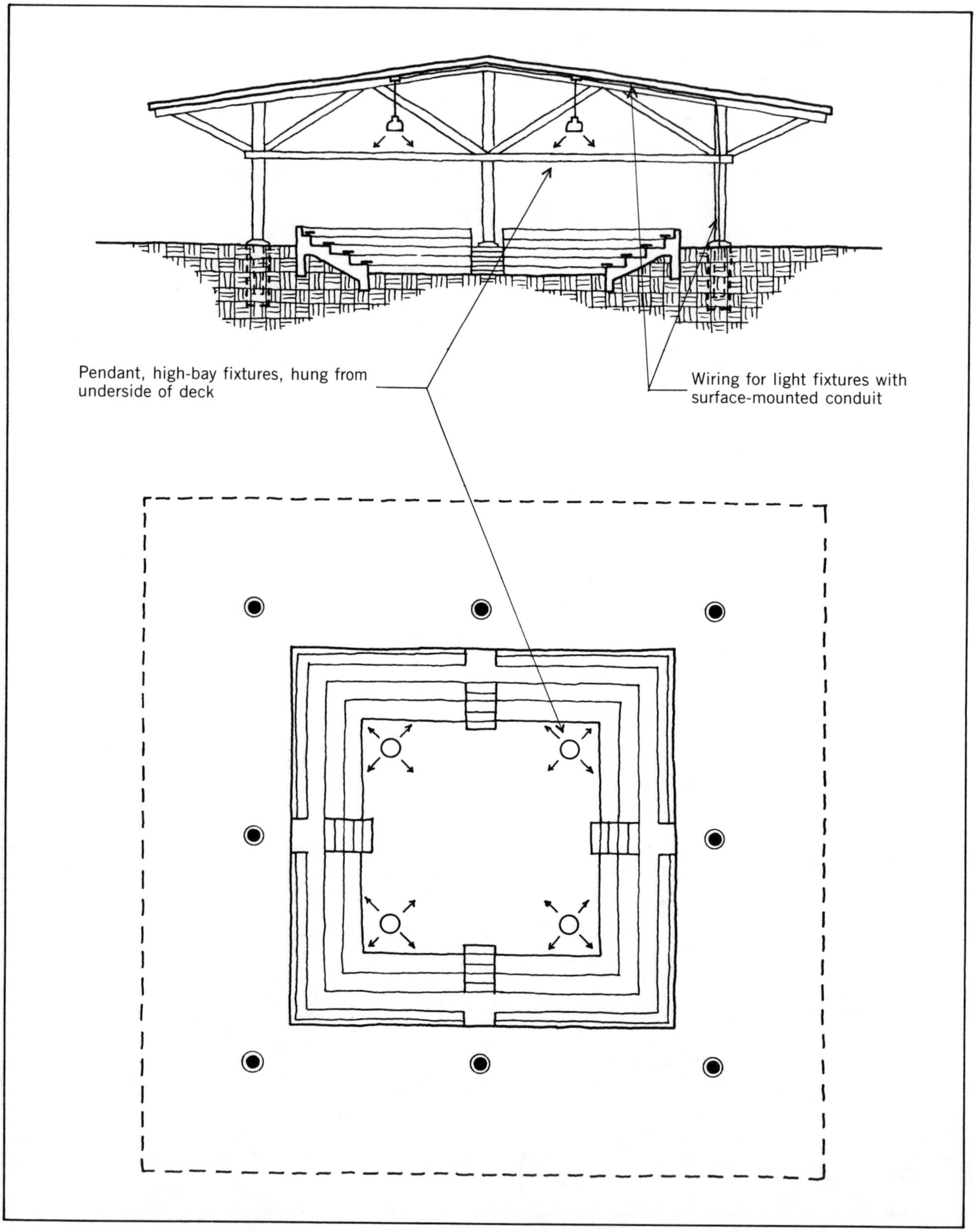

Figure 5.47 Building 9: details for the lighting system—option 1.

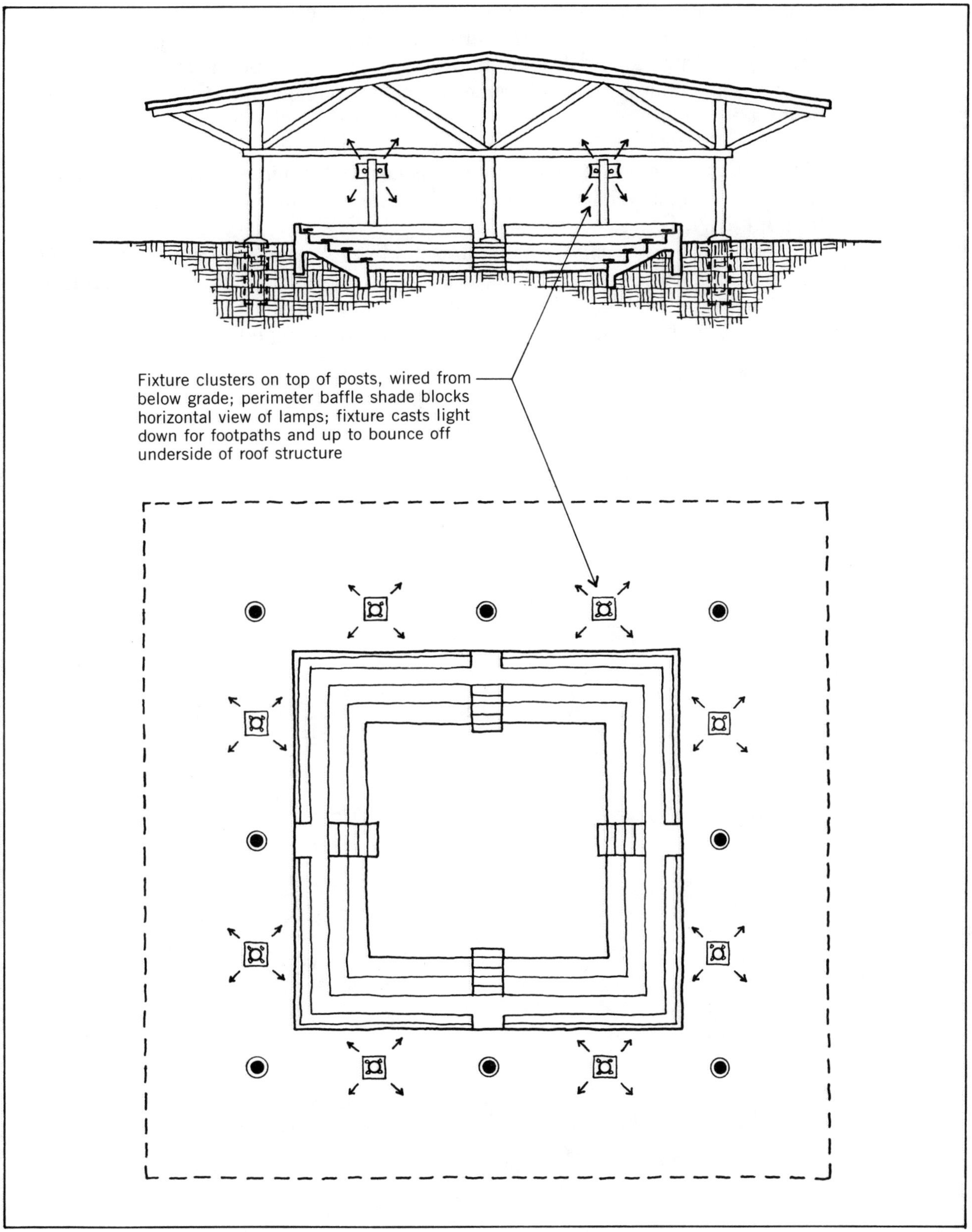

Figure 5.48 Building 9: details for the lighting system—option 2.

duits. With no other vertical elements, the surfact conduit would need to run up the face of one selected column. An inelegant, practical solution.

A slightly more graceful solution for lighting is shown in Figure 5.48. This system stays independent of the roof structure, using pole elements at the perimeter of the seating. The fixtures used in this case throw their light mostly up and down, not sideways. Fixtures on the opposite side would thus be less disturbing as glare spots to viewers. The downward-cast light will generally illuminate the seating, while the upward-cast light will light the underside of the roof and bounce some light into the amphitheater.

Depending on the site profile and groundwater conditions in general, it may be necessary to provide some drainage for the depressed amphitheater area. A leaching tile system under the amphitheater floor could be connected to a sewer or simply carried somewhere lower outside the building and dumped.

Index